Lecture Notes in Mathematics

Edited by J.-M. Morel, F. Takens and B. Teissier

Editorial Policy
for the publication of monographs

1. Lecture Notes aim to report new developments in all areas of mathematics and their applications- quickly, informally and at a high level. Mathematical texts analysing new developments in modelling and numerical simulation are welcome.

   Monograph manuscripts should be reasonably self-contained and rounded off. Thus they may, and often will, present not only results of the author but also related work by other people. They may be based on specialised lecture courses. Furthermore, the manuscripts should provide sufficient motivation, examples and applications. This clearly distinguishes Lecture Notes from journal articles or technical reports which normally are very concise. Articles intended for a journal but too long to be accepted by most journals, usually do not have this "lecture notes" character. For similar reasons it is unusual for doctoral theses to be accepted for the Lecture Notes series, though habilitation theses may be appropriate.

2. Manuscripts should be submitted (preferably in duplicate) either to Springer's mathematics editorial in Heidelberg, or to one of the series editors (with a copy to Springer). In general, manuscripts will be sent out to 2 external referees for evaluation. If a decision cannot yet be reached on the basis of the first 2 reports, further referees may be contacted: The author will be informed of this. A final decision to publish can be made only on the basis of the complete manuscript, however a refereeing process leading to a preliminary decision can be based on a pre-final or incomplete manuscript. The strict minimum amount of material that will be considered should include a detailed outline describing the planned contents of each chapter, a bibliography and several sample chapters.

   Authors should be aware that incomplete or insufficiently close to final manuscripts almost always result in longer refereeing times and nevertheless unclear referees' recommendations, making further refereeing of a final draft necessary.

   Authors should also be aware that parallel submission of their manuscript to another publisher while under consideration for LNM will in general lead to immediate rejection.

3. Manuscripts should in general be submitted in English. Final manuscripts should contain at least 100 pages of mathematical text and should always include

   - a table of contents;
   - an informative introduction, with adequate motivation and perhaps some historical remarks: it should be accessible to a reader not intimately familiar with the topic treated;
   - a subject index: as a rule this is genuinely helpful for the reader.

Continued on inside back-cover

Lecture Notes in Mathematics 1860

Alla Borisyuk   Avner Friedman
Bard Ermentrout   David Terman

Tutorials in
Mathematical Biosciences I

Mathematical Neuroscience

 Springer

Authors

Alla Borisyuk
Mathematical Biosciences Institute
The Ohio State University
231 West 18th Ave.
Columbus, OH 43210-1174, USA
e-mail: borisyuk@mbi.osu.edu

Avner Friedman
Mathematical Biosciences Institute
The Ohio State University
231 West 18th Ave.
Columbus, OH 43210-1174, USA
e-mail: afriedman@mbi.osu.edu

Bard Ermentrout
Department of Mathematics
University of Pittsburgh
502 Thackeray Hall
Pittsburgh, PA 15260, USA
e-mail: bard@pitt.edu

David Terman
Department of Mathematics
The Ohio State University
231 West 18th Ave.
Columbus, OH 43210-1174, USA
e-mail: terman@math.ohio-state.edu

Cover Figure: Cortical neurons (nerve cells), © Dennis Kunkel Microscopy, Inc.

Library of Congress Control Number: 2004117383

Mathematics Subject Classification (2000): 34C10, 34C15, 34C23, 34C25, 34C37, 34C55, 35K57, 35Q80, 37N25, 92C20, 92C37

ISSN 0075-8434
ISBN 3-540-23858-1 Springer-Verlag Berlin Heidelberg New York
DOI 10.1007/b102786

Springer is part of Springer Science+Business Media
springeronline.com

© Springer-Verlag Berlin Heidelberg 2005
Printed in Germany

Typesetting: Camera-ready TeX output by the authors

SPIN: 11348290   41/3142-543210 - Printed on acid-free paper

Preface

This is the first volume in the series "Tutorials in Mathematical Biosciences". These lectures are based on material which was presented in tutorials or developed by visitors and postdoctoral fellows of the Mathematical Biosciences Institute (MBI), at The Ohio State University. The aim of this series is to introduce graduate students and researchers with just a little background in either mathematics or biology to mathematical modeling of biological processes. The first volume is devoted to Mathematical Neuroscience, which was the focus of the MBI program in 2002-2003; documentation of this year's activities, including streaming videos of the workshops, can be found on the website http://mbi.osu.edu.

The use of mathematics in studying the brain has had great impact on the field of neuroscience and, simultaneously, motivated important research in mathematics. The Hodgkin-Huxley model, which originated in the early 1950s, has been fundamental in our understanding of the propagation of electrical impulses along a nerve axon. Reciprocally, the analysis of these equations has resulted in the development of sophisticated mathematical techniques in the fields of partial differential equations and dynamical systems. Interaction among neurons by means of their synaptic terminals has led to a study of coupled systems of ordinary differential and integro-differential equations, and the field of computational neurosciences can now be considered a mature discipline.

The present volume introduces some basic theory of computational neuroscience. Chapter 2, by David Terman, is a self-contained introduction to dynamical systems and bifurcation theory, oriented toward neuronal dynamics. The theory is illustrated with a model of Parkinson's disease. Chapter 3, by Bard Ermentrout, reviews the theory of coupled neural oscillations. Oscillations are observed throughout the nervous systems at all levels, from single cell to large network: This chapter describes how oscillations arise, what pattern they may take, and how they depend on excitory or inhibitory synaptic connections. Chapter 4 specializes to one particular neuronal system, namely, the auditory system. In this chapter, Alla Borisyuk provides a self-contained introduction to the auditory system, from the anatomy and physiology of the inner ear to the neuronal network which connects the hair cells to the cortex. She describes various models of subsystems such as the one that underlies sound localization. In Chapter 1, I have given a brief introduction to neurons, tailored to the subsequent chapters. In particular, I have included the electric circuit theory used to model the propagation of the action potential along an axon.

I wish to express my appreciation and thanks to David Terman, Bard Ermentrout, and Alla Borisyuk for their marvelous contributions. I hope this volume will serve as a useful introduction to those who want to learn about the important and exciting discipline of Computational Neuroscience.

August 27, 2004   Avner Friedman, Director, MBI

Contents

Introduction to Neurons
Avner Friedman 1
1 The Structure of Cells 1
2 Nerve Cells 6
3 Electrical Circuits and the Hodgkin-Huxley Model 9
4 The Cable Equation 15
References 20

An Introduction to Dynamical Systems
and Neuronal Dynamics
David Terman 21
1 Introduction 21
2 One Dimensional Equations 23
   2.1 The Geometric Approach 23
   2.2 Bifurcations 24
   2.3 Bistability and Hysteresis 26
3 Two Dimensional Systems 28
   3.1 The Phase Plane 28
   3.2 An Example 29
   3.3 Oscillations 31
   3.4 Local Bifurcations 31
   3.5 Global Bifurcations 33
   3.6 Geometric Singular Perturbation Theory 34
4 Single Neurons 36
   4.1 Some Biology 37
   4.2 The Hodgkin-Huxley Equations 38
   4.3 Reduced Models 39
   4.4 Bursting Oscillations 43
   4.5 Traveling Wave Solutions 47

5 Two Mutually Coupled Cells 50
   5.1 Introduction 50
   5.2 Synaptic Coupling 50
   5.3 Geometric Approach 51
   5.4 Synchrony with Excitatory Synapses 53
   5.5 Desynchrony with Inhibitory Synapses 57
6 Activity Patterns in the Basal Ganglia 61
   6.1 Introduction 61
   6.2 The Basal Ganglia 61
   6.3 The Model 62
   6.4 Activity Patterns 63
   6.5 Concluding Remarks 65
References 66

Neural Oscillators
Bard Ermentrout 69
1 Introduction 69
2 How Does Rhythmicity Arise 70
3 Phase-Resetting and Coupling Through Maps 73
4 Doublets, Delays, and More Maps 78
5 Averaging and Phase Models 80
   5.1 Local Arrays 84
6 Neural Networks 91
   6.1 Slow Synapses 91
   6.2 Analysis of the Reduced Model 94
   6.3 Spatial Models 96
References 103

**Physiology and Mathematical Modeling
of the Auditory System**
Alla Borisyuk 107
1 Introduction 107
   1.1 Auditory System at a Glance 108
   1.2 Sound Characteristics 110
2 Peripheral Auditory System 113
   2.1 Outer Ear 113
   2.2 Middle Ear 114
   2.3 Inner Ear. Cochlea. Hair Cells. 115
   2.4 Mathematical Modeling of the Peripheral Auditory System 117
3 Auditory Nerve (AN) 124
   3.1 AN Structure 124
   3.2 Response Properties 124
   3.3 How Is AN Activity Used by Brain? 127
   3.4 Modeling of the Auditory Nerve 130

4 Cochlear Nuclei 130
   4.1 Basic Features of the CN Structure 131
   4.2 Innervation by the Auditory Nerve Fibers 132
   4.3 Main CN Output Targets 133
   4.4 Classifications of Cells in the CN 134
   4.5 Properties of Main Cell Types 135
   4.6 Modeling of the Cochlear Nuclei 141
5 Superior Olive. Sound Localization, Jeffress Model 142
   5.1 Medial Nucleus of the Trapezoid Body (MNTB) 142
   5.2 Lateral Superior Olivary Nucleus (LSO) 143
   5.3 Medial Superior Olivary Nucleus (MSO) 143
   5.4 Sound Localization. Coincidence Detector Model 144
6 Midbrain 150
   6.1 Cellular Organization and Physiology of Mammalian IC 151
   6.2 Modeling of the IPD Sensitivity in the Inferior Colliculus 151
7 Thalamus and Cortex 161
References 162

Index 169

Introduction to Neurons

Avner Friedman

Mathematical Biosciences Institute, The Ohio State University, W. 18th Avenue 231, 43210-1292 Ohio, USA
afriedman@mbi.osu.edu

Summary. All living animals obtain information from their environment through sensory receptors, and this information is transformed to their brain where it is processed into perceptions and commands. All these tasks are performed by a system of nerve cells, or neurons. Neurons have four morphologically defined regions: the cell body, dendrites, axon, and presynaptic terminals. A bipolar *neuron* receives signals from the dendritic system; these signals are integrated at a specific location in the cell body and then sent out by means of the axon to the presynaptic terminals. There are neurons which have more than one set of dendritic systems, or more than one axon, thus enabling them to perform simultaneously multiple tasks; they are called *multipolar neurons*.

This chapter is not meant to be a text book introduction to the general theory of neuroscience; it is rather a brief introduction to neurons tailored to the subsequent chapters, which deal with various mathematical models of neuronal activities. We shall describe the structure of a generic bipolar neuron and introduce standard mathematical models of signal transduction performed by neurons. Since neurons are cells, we shall begin with a brief introduction to cells.

1 The Structure of Cells

Cells are the basic units of life. A cell consists of a concentrated aqueous solution of chemicals and is capable of replicating itself by growing and dividing. The simplest form of life is a single cell, such as a yeast, an amoeba, or a bacterium. Cells that have a nucleus are called *eukaryotes*, and cells that do not have a nucleus are called *prokaryotes*. Bacteria are prokaryotes, while yeasts and amoebas are eukaryotes. Animals are multi-cellular creatures with eukaryotic cells. A typical size of a cell is 5–20μm (1μm = 1 micrometer = 10^{-6} meter) in diameter, but an oocyte may be as large as 1mm in diameter. The human body is estimated to have 1014 cells. Cells may be very diverse as they perform different tasks within the body. However, all eukaryotic cells have the same basic structure composed of a nucleus, a variety of organelles and molecules, and a *plasma membrane*, as indicated in Figure 1 (an exception are the red blood cells, which have no nucleus).

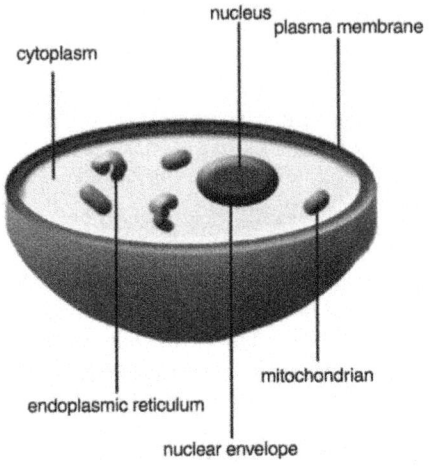

Fig. 1. A cell with nucleus and some organelles.

The DNA, the genetic code of the cell, consists of two strands of polymer chains having a double helix configuration, with repeated nucleotide units A, C, G, and T. Each A on one strand is bonded to T on the other strand by a hydrogen bond, and similarly each C is hydrogen bonded to T. The DNA is packed in chromosomes in the nucleus. In humans, the number of chromosomes in a cell is 46, except in the sperm and egg cells where their number is 23. The total number of DNA base pairs in human cells is 3 billions. The nucleus is enclosed by the nuclear envelope, formed by two concentric membranes. The nuclear envelope is perforated by *nuclear pores*, which allow some molecules to cross from one side to another.

The cell's plasma membrane consists of a lipid bilayer with proteins embedded in them, as shown in Figure 2. The *cytoplasm* is the portion of the cell which lies outside the nucleus and inside the cell's membrane.

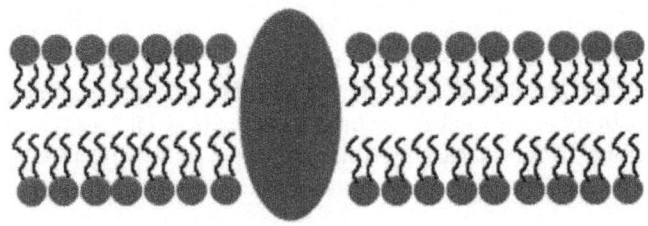

Fig. 2. A section of the cell's membrane.

An *organelle* is a discrete structure in the cytoplasm specialized to carry out a particular function. A *mitochondrion* is a membrane-delineated organelle that uses oxygen to produce energy, which the cell requires to perform its various tasks. An *endoplasmic reticulum* (ER) is another membrane-bounded organelle where lipids are secreted and membrane-bound proteins are made. The cytoplasm contains a number of mitochondria and ER organelles, as well as other organelles, such as *lysosomes* in which intra-cellular digestion occurs. Other structures made up of proteins can be found in the cell, such as a variety of filaments, some of which serve to strengthen the cell mechanically. The cell also contains amino acid molecules, the building blocks of proteins, and many other molecules.

The *cytoskeleton* is an intricate network of protein filaments that extends throughout the cytoplasm of the cell. It includes families of *intermediate filaments*, *microtubules*, and *actin filaments*. Intermediate filaments are rope-like fibers with a diameter of 10nm and strong tensile strength (1nm=1 nanometer=10^{-9} meter). Microtubules are long, rigid, hollow cylinders of outer diameter 25nm. Actin filaments, with diameter 7nm, are organized into a variety of linear bundles; they are essential for all cell movement such as crawling, engulfing of large particles, or dividing. Microtubules are used as a "railroad tract" in transport of vesicles across the cytoplasm by means of motor proteins (see next paragraph). The motor protein has one end attached to the vescicle and the other end, which consists of two "heads", attached to the microtubule. Given input of energy, the protein's heads change configuration (conformation), thereby executing one step with each unit of energy.

Proteins are polymers of amino acids units joined together head-to-tail in a long chain, typically of several hundred amino acids. The linkage is by a covalent bond, and is called a *peptide bond*. A chain of amino acids is known as a *polypeptide*. Each protein assumes a 3-dimensional configuration, which is called a *conformation*. There are altogether 20 different amino acids from which all proteins are made. Proteins perform specific tasks by changing their conformation.

The various tasks the cell needs to perform are executed by proteins. Proteins are continuously created and degraded in the cell. The synthesis of proteins is an intricate process. The DNA contains the genetic code of the cell. Each group of three letters (or three base pairs) may be viewed as one "word". Some collections of words on the DNA represent *genes*. The cell expresses some of these genes into proteins. This translation process is carried out by several types of RNAs: *messenger* RNA (mRNA), *transfer* RNA (tRNA), and *ribosomal* RNA (rRNA). Ribosome is a large complex molecule made of more than 50 different ribosomal proteins, and it is there where proteins are synthesized. When a new protein needs to be made, a signal is sent to the DNA (by a *promoter* protein) to begin transcribing a segment of a strand containing an appropriate gene; this copy of the DNA strand is the mRNA. The mRNA molecule travels from the nucleus to a ribosome, where each "word" of three letters, for example (A, C, T), called a *codon*, is going to be translated into one amino acid. The translation is accomplished by tRNA, which is a relatively compact molecule. The tRNA has a shape that is particularly suited to conform to the codon at one end and is attached to an amino acid corresponding to the particular codon at its other end. Step-by-step, or one-by-one, the tRNAs line up along the ribosome, one codon at a time, and at each step a new amino acid is brought in to the ribosome where it connects to the preceding amino acid, thus joining the growing chain of amino acids until the entire protein is synthesized.

The human genome has approximately 30,000 genes. The number of different proteins is even larger; however cells do not generally express all their genes.

The cell's membrane is typically 6–8nm thick and as we said before, it is made of a double layer of lipids with proteins embedded throughout. The lipid bilayer is hydrophobic and selectively permeable. Small nonpolar molecules such as O_2 and CO_2 readily dissolve in the lipid bilayer and rapidly diffuse across it. Small uncharged polar molecules such as water and ethanol also diffuse rapidly across the bilayer. However, larger molecules or any ions or charged molecules cannot diffuse across the lipid bilayer. These can only be selectively transported across the membrane by proteins, which are embedded in the membrane. There are two classes of such proteins: *carrier proteins* and *channel proteins*. Carrier proteins bind to a solute on one side of the membrane and then deliver it to the other side by means of a change in their conformation. Carrier proteins enable the passage of nutrients and amino acids into the cell, and the release of waste products, into the extracellular environment. Channel proteins form tiny hydrophilic pores in the membrane through which the solute can pass by diffusion. Most of the channel proteins let through only inorganic ions, and these are called *ion channels*.

Both the intracellular and extracellular environments include ionized aqueous solution of dissolved salts, primarily $NaCl$ and KCl, which in their disassociated state are Na^+, K^+, and Cl^- ions. The concentration of these ions, as well as other ions such as $Ca2^+$, inside the cell differs from their concentration outside the cell. The concentration of Na^+ and $Ca2^+$ inside the cell is smaller than their concentration outside the cell, while K^+ has a larger concentration inside the cell than outside it. Molecules move from high concentration to low concentration ("downhill" movement). A pathway that is open to this movement is called a *passive* channel or a *leak* channel; it does not require expenditure of energy. An *active transport* channel is one that transports a solute from low concentration to high concentration ("uphill" movement); such a channel requires expenditure of energy.

An example of an active transport is the sodium-potassium pump, pumping $3Na^+$ out and $2K^+$ in. The corresponding chemical reaction is described by the equation

$$ATP + 3Na_i^+ + 2K_e^+ \rightarrow ADP + P_i + 3Na_e^+ + 2K_i^+$$

In this process, energy is expended by the conversion of one molecule ATP to one ADP and a phosphate atom P.

Another example of active transport is the calcium pump. The concentration of free $Ca2^+$ in the cell is 0.1μM, while the concentration of Ca^{2+} outside the cell is 1mM, that is, higher by a factor of 10^4 (μM=micromole=10^{-6} mole, mM=milimole=10^{-3} mole, mole=number of grams equal to the molecular weight of a molecule). To help maintain these levels of concentration the cell uses active calcium pumps.

An important formula in electrophysiology and in neuroscience is the *Nernst equation*. Suppose two reservoirs of the same ions S with, say, a positive charge Z per ion, are separated by a membrane. Suppose each reservoir is constantly kept electrically neutral by the presence of other ions T. Finally, suppose that the membrane is permeable to S but not to T. We shall denote by $[S_i]$ the concentration of S on the left side or the inner side, of the membrane, and by $[S_o]$ the concentration of S on the right side, or the outer side, of the membrane. If the concentration $[S_i]$ is initially larger than the concentration $[S_o]$, then ions S will flow from inside the membrane to the outside, building up a positive charge that will increasingly resist further movement of positive ions from the inside to the outside of the membrane. When equilibrium is reached, $[S_o]$ will be, of course, larger than $[S_i]$ and (even though each side of the membrane is electrically neutral) there will be voltage difference V_s across the membrane. V_s is given by the Nernst equation

$$V_s = \frac{RT}{ZF} \ell n \frac{[S_o]}{[S_i]}$$

when R is the universal gas constant, F is the Faraday constant, and T is the absolute temperature. For $Z = 1$, temperature=37°C,

$$V_s = 62 \log_{10} \frac{[S_o]}{[S_i]} \ .$$

By convention, the membrane potential is defined as the difference: The outward-pointing electric field from inside the membrane minus the inward-pointing electric field from outside the membrane.

The ions species separated by the cell membrane, are primarily K^+, Na^+, Cl^-, and Ca^{2+}. To each of them corresponds a different Nernst potential. The electric potential at which the net electrical current is zero is called the *resting membrane potential*. An approximate formula for computing the resting membrane potential is known as the Goldman-Hodgkin-Katz (GHK) equation.

For a typical mammalian cell at temperature 37°C,

S	$[S_i]$	$[S_o]$	V_s
K^+	140	5	−89.7 mV
Na^+	5–15	145	+90.7 − (+61.1)mV
Cl^-	4	110	−89mV
Ca^{2+}	1–2	2.5–5	+136 − (+145)mV

where the concentration is in milimolar (mM) and the potential is in milivolt. The negative V_s for $S = K^+$ results in an inward-pointing electric field which drives the positively charged K^+ ions to flow inward. The sodium-potassium pump is used to maintain the membrane potential and, consequently, to regulate the cell volume. Indeed, recall that the plasma membrane is permeable to water. If the total concentration of solutes is low one side of the membrane and high on the other, then water will tend to move across the membrane to make the solute concentration equal; this process is known as *osmosis*. The osmotic pressure, which drives water across the cell, will cause the cell to swell and eventually to burst, unless it is countered by an equivalent force, and this force is provided by the membrane potential. The resting potential for mammalian cells is in the range of -60mV to -70mV.

2 Nerve Cells

There are many types of cells in the human body. These include: (i) a variety of epithelial cells that line up the inner and outer surfaces of the body; (ii) a variety of cells in connective tissues such as fibroblasts (secreting extracellular protein, such as collagen and elastin) and lipid cells; (iii) a variety of muscle cells; (iv) red blood cells and several types of white blood cells; (v) sensory cells, for example, rod cells in the retina and hair cells in the inner ear; and (vi) a variety of nerve cells, or neurons.

The fundamental task of neurons is to receive, conduct, and transmit signals. Neurons carry signals from the sense organs inward to the central nervous system (CNS), which consists of the brain and spinal cord. In the CNS the signals are analyzed and interpreted by a system of neurons, which then produce a response. The response is sent, again by neurons, outward for action to muscle cells and glands.

Neurons come in many shapes and sizes, but they all have some common features as shown schematically in Figure 3.

A typical neuron consists of four parts: *cell body*, or *soma*, containing the nucleus and other organelles (such as ER and mitochondria); branches of *dendrites*, which receive signals from other neurons; an *axon* which conducts signals away from the cell body; and many branches at the far end of the axon, known as *nerve terminals* or *presynaptic terminals*. Nerve cells, body and axon, are surrounded by *glial cells*. These provide support for nerve cells, and they also provide insulation sheaths called *myelin* that cover and protect most of the large axons. The combined number of neurons and glial cells in the human body is estimated at 10^{12}.

The length of an axon varies from less than 1mm to 1 meter, depending on the type of nerve cell, and its diameter varies between 0.1μm and 20μm.

The dendrites receive signals from nerve terminals of other neurons. These signals, tiny electric pulses, arrive at a location in the soma, called the *axon hillock*. The combined electrical stimulus at the hillock, if exceeding a certain

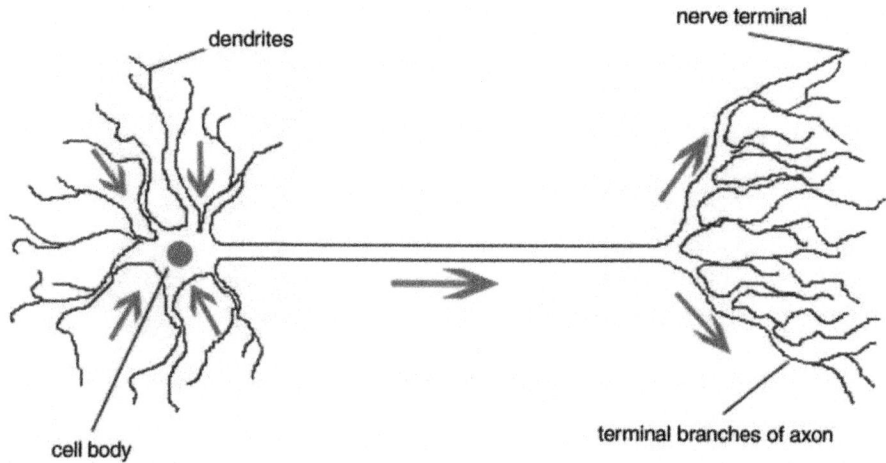

Fig. 3. A neuron. The arrows indicate direction of signal conduction.

threshold, triggers the initiation of a traveling wave of electrical excitation in the plasma membrane known as the *action potential*. If the plasma membrane were an ordinary conductor, then the electrical pulse of the action potential would weaken substantially along the plasma membrane. However, as we shall see, the plasma membrane, with its many sodium and potassium active channels spread over the axon membrane, is a complex medium with conductance and resistance properties that enable the traveling wave of an electrical excitation to maintain its pulse along the plasma membrane of the axon without signal weakening. The traveling wave has a speed of up to 100m/s.

A decrease in the membrane potential (for example, from -65mV to -55mV) is called *depolarization*. An increase in the membrane potential (for example, from -65mV to -75mV) is called *hyperpolarization*. Depolarization occurs when a current is injected into the plasma membrane. As we shall see, depolarization enables the action potential, whereas hyperpolarization tends to block it. Hence, a depolarizing signal is *excitatory* and a hyperpolarizing signal is *inhibitory*.

The action potential is triggered by a sudden depolarization of the plasma membrane, that is, by a shift of the membrane potential to a less negative value. This is caused in many cases by ionic current, which results from stimuli by neurotransmitters released to the dendrites from other neurons. When the depolarization reaches a threshold level (e.g., from -65mV to -55mV) it affects voltage-gated channels in the plasma membrane. First, the sodium channels at the site open: the electrical potential difference across the membrane causes conformation change, as illustrated in Figure 4, which results in the opening of these channels.

When the sodium channels open, the higher Na^+ concentration on the outside of the axon pressures these ions to move into the axon against the

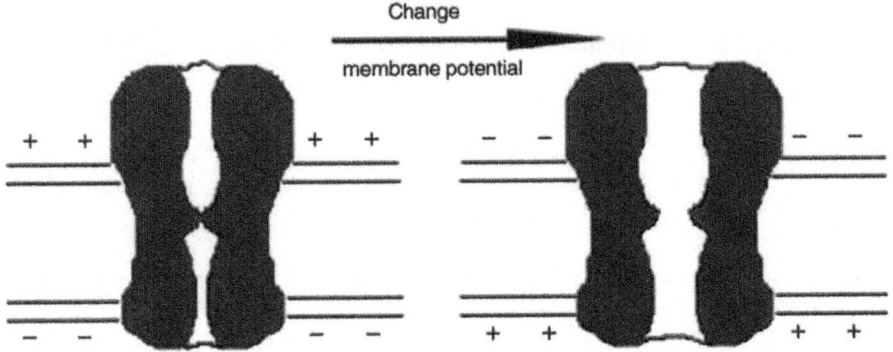

Fig. 4. Change in membrane voltage can open some channels.

depolarized voltage; thus, the sodium ions flow from outside to inside the axon along the electrochemical gradient. They do so at the rate of 10^8 ions per second. This flow of positive ions into the axon further enhances the depolarization, so that the voltage V_m of the plasma membrane continues to increase.

As the voltage continues to increase (but still being negative), the potassium channels at the site begin to open up, enabling K^+ ions to flow out along the electrochemical gradient. However, as long as most of the sodium channels are still open, the voltage nevertheless continues to increase, but soon the sodium channels shut down and, in fact, they remain shut down for a period of time called the *refractory period.*

While the sodium channels are in their refractory period, the potassium channels remain open so that the membrane potential (which arises, typically, to +50mV) begins to decrease, eventually going down to its initial depolarized state where again new sodium channels, at the advanced position of the action potential, begin to open, followed by potassium channels, etc. In this way, step-by-step, the action potential moves along the plasma membrane without undergoing significant weakening. Figure 5 illustrates one step in the propagation of the action potential.

Most ion channels allow only one species of ions to pass through. Sodium channels are the first to open up when depolarization occurs; potassium channels open later, as the plasma potential is increased. The flux of ions through the ion channels is passive; it requires no expenditure of energy. In addition to the flow of sodium and potassium ions through voltage-gated channels, transport of ions across the membrane takes place also outside the voltage-gated channels. Indeed, most membranes at rest are permeable to K^+, and to a (much) lesser degree to Na^+ and Ca^{2+}.

As the action potential arrives at the nerve terminal, it transmits a signal to the next cell, which may be another neuron or a muscle cell. The spacing through which this signal is transmitted is called the *synaptic cleft.* It

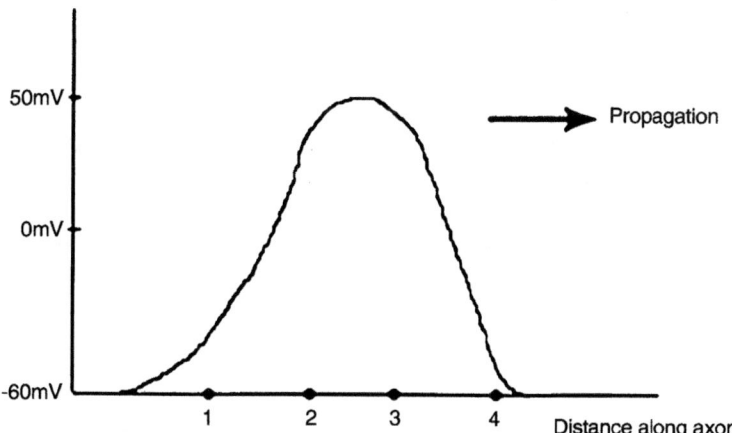

Fig. 5. Propagation of the action potantial. 1: Na+ channels open; 2: K+ channels open; 3: Na+ channels close; 4: k+ channels close.

separates the *presynaptic* cytoplasm of the neuron from the *postsynaptic* cell. There are two types of synaptic transmissions: chemical and electrical. Figure 6 shows a chemical synaptic transmission. This involves several steps: The action potential arriving at the presynaptic axon causes voltage-gated Ca^{2+} channels near the synaptic end to open up. Calcium ions begin to flow into the presynaptic region and cause vesicles containing neurotransmitters to fuse with the cytoplasmic membrane and release their content into the synaptic cleft. The released neurotransmitters diffuse across the synaptic cleft and bind to specific protein receptors on the postsynaptic membrane, triggering them to open (or close) channels, thereby changing the membrane potential to a depolarizing (or a hyperpolarizing) state. Subsequently, the neurotransmitters recycle back into their presynaptic vesicles.

Electrical transmission is when the action potential makes direct electrical contact with the postsynaptic cell. The gap junction in electrical transmission is very narrow; about 3.5nm. Chemical transmission incurs time delay and some variability due to the associated diffusion processes, it requires a threshold of the action potential, and it is unidirectional. By contrast, electrical transmission incurs no time delay, no variability, it requires no threshold, and it is bidirectional between two neurons.

3 Electrical Circuits and the Hodgkin-Huxley Model

The propagation of the action potential along the axon membrane can be modeled as the propagation of voltage in an electrical circuit. Before describing this model, let us review the basic theory of electrical circuits. We begin with

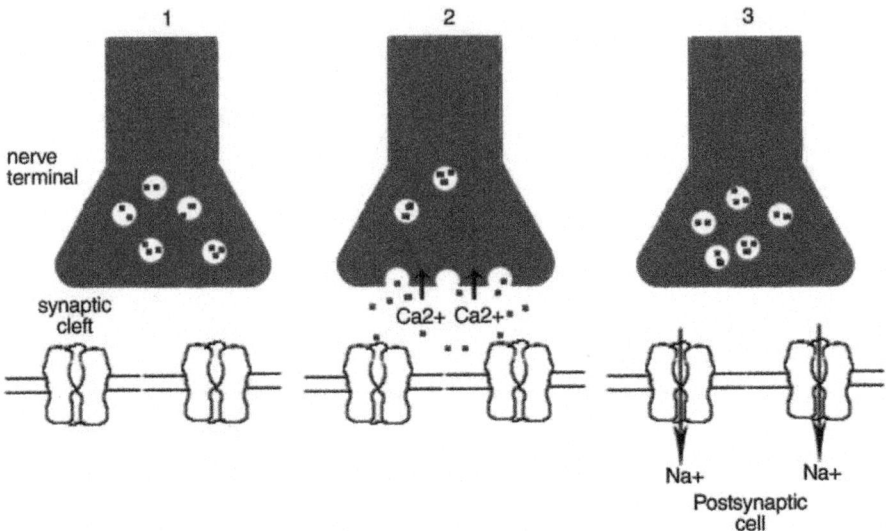

Fig. 6. Synaptic transmission at chemical synapses. 1: Arrival of action potential. 2: Ca^{2+} flows in; vesicles fuse to cytoplasm membrane, and release their contents to the synaptic cleft. 3: Postsynaptic (e.g. Na^+) channels open, and Ca^{2+} ions return to vesicles.

the *Coulomb law*, which states that positive charges q_1 and q_2 at distance r from each other experience a repulsive force F given by

$$F = \frac{1}{4\pi\varepsilon_o} \frac{q_1 q_2}{r^2}$$

where ε_0 is the permittivity of space. We need of course to define the unit of charge, C, called *coulomb*. A coulomb, C, is a quantity of charge that repels an identical charge situated 1 meter away with force $F = 9 \times 10^9 N$, where N=newton=10^5 dyne. This definition of C is clearly related to the value of ε_0, which is such that

$$\frac{1}{4\pi\varepsilon_o} = 9 \times 10^9 \frac{Nm^2}{C^2}$$

The charge of an electron is $-e$, where $e = 1.602 \times 10^{-19}C$. Hence the charge of one mole of ions K^+, or of one mole of any species of positive ions with a unit charge e per ion, is $N_A C$ where $N_A = 6.023 \times 10^{23}$ is the *Avogadro number*. The quantity $F = N_A e = 96,495C$ is called the *Faraday constant*.

Electromotive force (EMF or, briefly, E) is measured in volts (V). One volt is the potential difference between two points that requires expenditure of 1 joule of work to move one coulomb of charge between the two points; 1 joule=10^7 erg=work done by a force of one Newton acting along a distance of 1 meter.

Current i is the rate of flow of electrical charge (q):

$$i = \frac{dq}{dt}.$$

Positive current is defined as the current flowing in the direction outward from the positive pole of a battery toward the negative pole; the electrons are then flowing in the opposite direction. In order to explain this definition, consider two copper wires dipped into a solution of copper sulfate and connected to the positive and negative poles of a battery. Then the positive copper ions in the solution are repelled from the positive wire and migrate toward the negative wire, whereas the negative sulfate ions move in the reverse direction. Since the direction of the current is defined as the direction from the positive pole to the negative pole, i.e., from the positive wire to the negative wire, the negative charge (i.e., the extra electrons of the surface atoms) move in the reverse direction.

The unit of current is *ampere*, A: One ampere is the flow of one coulomb C per second.

Ohm's law states that the ratio of voltage V to current I is a constant R, called the *resistance*:

$$R = \frac{V}{I}$$

R is measured in ohms, $\Omega : \Omega = \frac{1V}{1A}$. Conductors, which satisfy the ohm law are said to be *ohmic*. Actually not all conductors satisfy Ohm's law; most neurons are *nonohmic* since the relation I–V is nonlinear. The quantity $\frac{1}{R}$ is called the *conductivity* of the conductor.

Capacitance is the ability of a unit in an electric circuit, called *capacitor*, to store charge; *capacity* C is defined by

$$C = \frac{q}{V}(C = \text{capacity}).$$

Where q is the stored charge and V is the potential difference (voltage) across the capacitor.

The unit capacity is *Farad, F*:

$$1F = \frac{1 coulomb}{1 volt}.$$

A typical capacitor consists of two conducting parallel plates with area S each, separated a distance r by a dielectric material with dielectric constant K_d. The capacity is given by

$$C = \varepsilon_o - K_d - \frac{S}{r}.$$

Later on we shall model a cell membrane as a capacitor with the bilipid layer as the dielectric material between the inner and outer surfaces of the plasma membrane.

It should be emphasized that no current ever flows across a capacitor (although the electric field force goes through it). However, when in an electric circuit the voltage changes in time, the charge on the capacitor also changes in time, so that it appears as if current is flowing. Since $i = \frac{dq}{dt}$ where $q = CV$ is the charge on the conductor, the *apparent* flow across the capacitor is

$$i = C\frac{dV}{dt}$$

(although there is no actual flow across it); we call this quantity the *capacitative current*. This flow merely reflects shifts of charge from one side of the capacitor to another by way of the circuit.

Kirchoff's laws form the basic theory of electrical circuits:

(1) The algebraic sum of all currents flowing toward a junction is zero; here, current is defined as positive if it flows into the junction and negative if it flows away from the junction.
(2) The algebraic sum of all potential sources and voltage drops passing through a closed conduction path (or "loop") is zero.

We give a simple application of Kirchoff's laws for the circuits described in Figures 7a and 7b. In figure 7a two resistors are in sequence, and Kirchoff's laws and Ohm's law give

$$E - 1R_1 - 1R_2 = 0 .$$

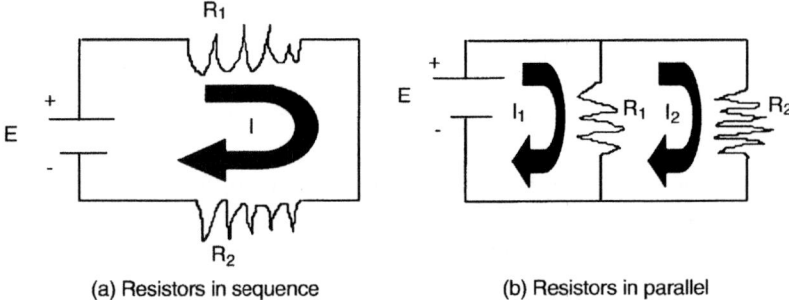

(a) Resistors in sequence   (b) Resistors in parallel

Fig. 7.

If the total resistance in the circuit is R, then also $E = IR$, by Ohm's law. Hence $R = R_1 + R_2$. By contrast, applying Kirchoff's law to the circuit described in Figure 7b, where the two resistances are in parallel, we get

$$I_1 = \frac{V}{R_1}, I_2 = \frac{V}{R_2}, \quad \text{and} \quad R = \frac{V}{I} \quad \text{where} \quad I = I_1 + I_2 ,$$

so that

$$\frac{1}{R} = \frac{1}{R_1} + \frac{1}{R_2}.$$

Capacitors introduce time element into the analysis of current flow. Indeed, since they accumulate and store electrical charge, current and voltage changes are no longer simultaneous. We shall illustrate this in an electric circuit, which resembles the cell membrane of an axon, as shown in Figure 8.

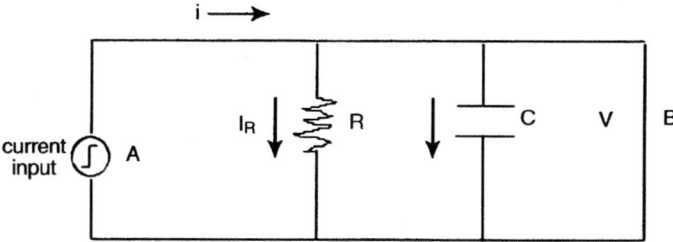

Fig. 8. Current step input.

On the left side A we introduce a current step i, as input, and on the right side B we measure the output voltage V; the capacitor C represents the capacity of the axon membrane and the resistor R represents the total resistivity of the ion channels in the axon. Thus, the upper line with the input $i \rightarrow$ may be viewed as the inner surface of the cell membrane, whereas the lower line represents the outer surface of the membrane.

By Kirchoff's laws and Ohm's law,

$$I_R = \frac{V}{R}, I_C = C\frac{dV}{dt}$$

and

$$iR = (I_R + I_C)R = V + RC\frac{dV}{dt}$$

so that

$$V = iR\left(1 - e^{-t/RC}\right).$$

Hence the voltage does not become instantly equal to iR (as it would be by Ohm's law if there was no capacitor in the circuit); V is initially equal to zero, and it increases to iR as $t \rightarrow \infty$.

Figure 9 describes a more refined electric circuit model of the axon membrane. It includes currents through potassium and sodium channels as well as a leak current, which may include Cl^- and other ion flows. For simplicity we have lumped all the channels of one type together as one channel and represented the lipid layer as a single capacitor; a more general model will be given in §4.

Since K^+ has a larger concentration inside the cell that outside the cell, we presume that positive potassium ions will flow outward, and we therefore denote the corresponding electromotive force E_K by

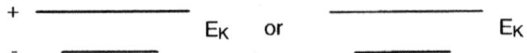

The reverse situation holds for Na^+. The conductivity of the channels K^+, Na^+ and the leak channel L are denoted by g_K, g_{Na}, and g_L, respectively.

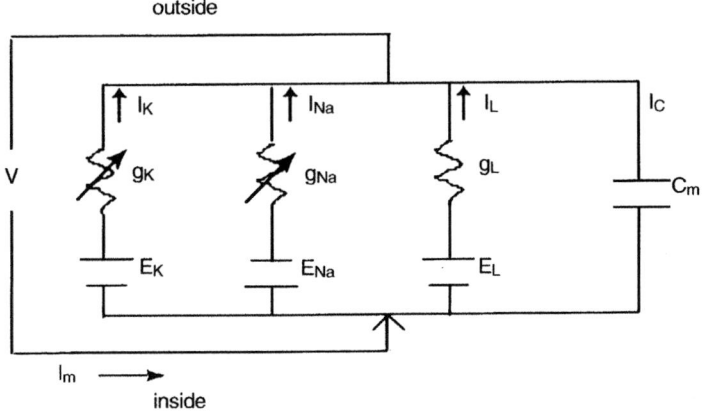

Fig. 9. An electric circuit representing an axon membrane.

By Kirchoff's laws we get

$$I_m = C_m \frac{dV}{dt} + I_K + I_{Na} + I_L$$

where V is the action potential and I_m is the current injected into the axon. We assume that the leak channel is ohmic, so that

$$I_L = g_L(V - E_L), g_L \text{ constant}$$

where E_L is the Nernst equilibrium voltage for this channel. On the other hand the conductivities g_K and g_{Na} are generally functions of V and t, as pointed out in §2, so that

$$I_K = g_K(V, t)(V - E_K), I_{Na} = g_{Na}(V, t)(V - E_{Na})$$

where E_K and E_{Na} are the Nernst equilibrium voltages for K^+ and Na^+. Thus, we can write

$$I_m = C_m \frac{dV}{dt} + g_K(V, t)(V - E_K) + g_{Na}(V, t)(V - E_{Na}) + g_L(V - E_L). \quad (1)$$

Hodgkin and Huxley made experiments on the giant squid axon, which consisted of clumping the voltage at different levels and measuring the corresponding currents. Making some assumptions on the structure and conformation of potassium and sodium gates, they proposed the following equations:

$$g_K(V, t) = n^4 g_K^*, g_{Na}(V, t) = m^3 h g_{Na}^* \tag{2}$$

where n, m, h are the *gating variables* (g_L is neglected here), and $g_K^* = \max g_K$, $g_{Na}^* = \max g_{Na}$. The variables n, m, h satisfy linear kinetic equations

$$\frac{dn}{dt} = \alpha_n(1 - n) - \beta_n n,$$

$$\frac{dm}{dt} = \alpha_m(1 - m) - \beta_m m,$$

$$\frac{dh}{dt} = \alpha_n(1 - h) - \beta_h h. \tag{3}$$

By fitting coefficients they obtained the empirical formulas

$$\alpha_n(V) = 0.01 \frac{-V + 10}{\left[e^{(-V+10)/10} - 1\right]},$$

$$\beta_n(V) = 0.125 e^{-V/80},$$

$$\alpha_m(V) = 0.1 \frac{-V + 25}{\left[e^{(-V+25)/10} - 1\right]},$$

$$\beta_m(V) = 4e^{-V/18}, \ \alpha_h(V) = 0.07 e^{-V/20}, \ \beta_h(V) = \frac{1}{e^{(-V+30)/10} + 1}. \tag{4}$$

The system (1)–(4) is known as the *Hodgkin-Huxley equations*. They form a system of four nonlinear ordinary differential equations in the variables V, n, m, and h. One would like to establish for this system, either by a mathematical proof or by computations, that as a result of a certain input of current I_m there will be solutions of (1)–(4) where the voltage V is, for example, a periodic function, or a traveling wave, as seen experimentally. This is an ongoing active area of research in the mathematical neuroscience.

The Hodgkin-Huxley equations model the giant squid axon. There are also models for other types of axons, some involving a smaller number of gating variables, which make them easier to analyze.

In the next section we shall extend the electric circuit model of the action potential to include distributions of channels across the entire axon membrane. In this case, the action potential will depend also on the distance x measured along the axis of the axon.

4 The Cable Equation

We model an axon as a thin cylinder with radius a. The interior of the cylinder (the cytoplasm) is an ionic medium which conducts electric current; we shall call it a *core conductor*. The exterior of the cylinder is also an ionic medium, and we shall assume for simplicity that it conducts current with no resistance. We introduce the following quantities:

r_i = axial resistance of the core conductor, $\frac{\Omega}{cm}$,

R_i = specific intercellular resistance, $\Omega \cdot cm$,

r_m = membrane resistance, $\Omega \cdot cm$,

R_m = specific membrane resistance, $\Omega \cdot cm^3$,

c_m = membrane capacitance, $\frac{F}{cm}$,

C_m = specific membrane capacitance, $\frac{F}{cm^3}$.

Then

$$R_i = \pi a^2 r_i, \; R_m = 2\pi a r_m, \; C_m = \frac{c_m}{2\pi a}; \tag{5}$$

the first equation, for example, follows by observing that r_i may be viewed as the resistance of a collection of resistances R_i in parallel.

Denote by x the distance along the axis of the core conductor, and by V_i the voltage in the core conductor. We assume that the current flows along the x-direction, so that V_i is a function of just (x,t). By Ohm's law

$$\frac{\partial V}{\partial x} = -i_i r_i .$$

Where i_i is the intracellular current; for definiteness we take the direction of the current flow to be in the direction of increase of x. Hence

$$\frac{\partial^2 V_i}{\partial x^2} = -r_i \frac{\partial i_i}{\partial x}. \tag{6}$$

If current flows out of (or into) the membrane over a length increment Δx then the current decreases (or increases) over that interval, as illustrated in Figure 10.

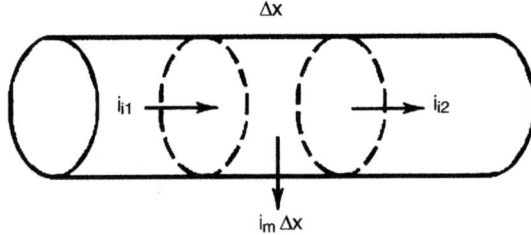

Fig. 10. Decrease in current i_i due to outflow of current i_m.

Denoting by i_m the flow, per unit length, out of (or into) the membrane, we have

$$i_2 - i_1 = -i_m \Delta x ,$$

or

$$i_m = -\frac{\partial i_i}{\partial x} .$$

Combining this with (6), we find that

$$\frac{1}{r_i}\frac{\partial^2 V_i}{\partial x^2} = i_m \tag{7}$$

The membrane potential is $V = V_i - V_e$ where V_e, the external voltage, is constant since we have assumed that the extracellular media has no resistance.

We now make the assumption that there are many identical circuits distributed all over the surface of the membrane, as illustrated in Figure 11.

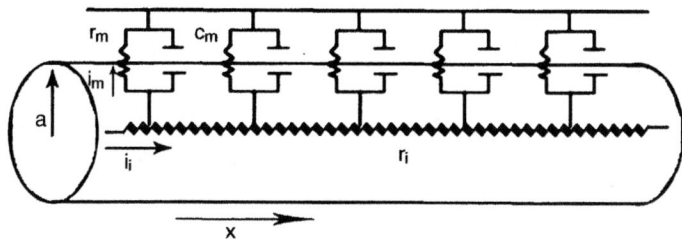

Fig. 11. Current flow along a cylindrical axon with many R-C circuits on the membrane.

By Kirchoff's laws (cf. Section 3) the flow i_m satisfies

$$i_m = c_m\frac{\partial V}{\partial t} + \frac{V_i - V_e}{r_m}. \tag{8}$$

Combining this with (7) we get

$$\frac{r_m}{r_i}\frac{\partial^2 V}{\partial x^2} = r_m c_m\frac{\partial V}{\partial t} + V.$$

Setting

$$\tau_m = r_m c_m = R_m C_m, \lambda = \sqrt{\frac{r_m}{r_i}} = \sqrt{\frac{R_m}{R_i}\frac{a}{2}},$$

we arrive at the *cable equation*

$$\lambda^2\frac{\partial^2 V}{\partial x^2} = \tau_m\frac{\partial V}{\partial t} + V, \tag{9}$$

or, with $X = \frac{x}{\lambda}, T = \frac{t}{\tau_m}$,

$$\frac{\partial V}{\partial t} = \frac{\partial^2 V}{\partial x^2} - V. \tag{10}$$

The specific current of the membrane, I_m, is related to the current i_m by $i_m = 2\pi a I_m$. Hence (7) can be written in the form

$$\frac{a}{2R_i}\frac{\partial^2 V}{\partial x^2} = I_m \ . \tag{11}$$

In the above analysis the current im was modeled using the configuration of R-C units as in Figure 11. Let us now specialize to the case of the giant squid axon, as illustrated in Figure 7, with the leak flow neglected. Then, as in §3,

$$I_m = C_m \frac{\partial V}{\partial t} + I_K + I_{Na}$$

so that, by (11),

$$\frac{a}{2R_i}\frac{\partial^2 V}{\partial x^2} = C_m \frac{\partial V}{\partial t} + g_K(V,t)(V - E_k) + g_{Na}(V,t)(V - E_{Na}) \tag{12}$$

where g_K, g_{Na} are as in the Hodgkin-Huxley equations. The *spatial Hodgkin-Huxley system* consists of the equations (2)–(4) and (12).

We return to the cable equation (9) and note that this equation can also model the voltage in any dendritic branch. But since the dendritic tree is quite complex in general (see Figure 3), it is difficult to the compute the total voltage, which sums up all the voltage inputs that go into the axon hillock. There is however one special case where V can be easily computed. This case was identified by Rall, and it assumes a very special relationship between some of the dendritic branches. In order to explain the *Rall Theory*, we begin with a situation in which a current I_0 is injected into an infinite core conductor (with $^-\infty < x < \infty$) at $x = 0$. Then current $I_0/2$ flows to the right and current $I_0/2$ flows to the left, so that the potential V satisifies

$$\frac{\partial V}{\partial x}(+0) - \frac{\partial V}{\partial x}(-0) = -r_i I_0 \ . \tag{13}$$

The stationary solution of the cable equation in the semi-infinite portion $0 < x < \infty$ is then

$$V(x) = r_i \frac{I_0}{2}\lambda e^{-x/\lambda} \ . \tag{14}$$

Note that the factor $e^{x-\lambda}$ accounts for the current leak through the membrane: If there is no leak then $V(x) = V(0)$. The resistance of the cable is then

$$V(0)\frac{I_0}{2} = \frac{(R_m R_i/2)^{1/2}}{\pi a^3} \ .$$

Hence the conductivity of the core conductor is

$$G = Kd^{3/2} \quad \text{where} \quad K = \frac{\pi}{2}(R_m R_i)^{-1} \tag{15}$$

where d is the diameter of the cable.

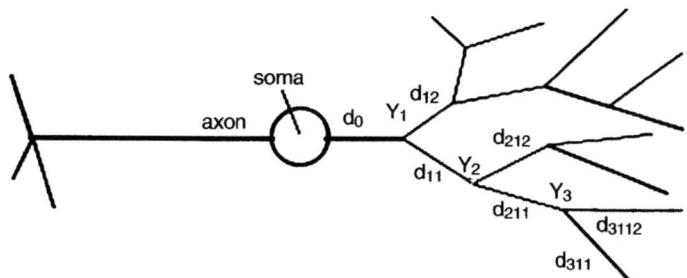

Fig. 12. Schematic diagram of a neuron with a branched dendritic tree.

Suppose the dendritic tree has a form as in Figure 12. We assume that each of end-branch is infinitely long, so that its conductivity is given by (15) where d is its diameter.

The conductances of the end-branches with diameters d_{3111} and d_{3112} are $K(d_{3111})^{3/2}$ and $Kd_{3112}^{3/2}$ respectively. Since the total conductances are just the sum of all the parallel conductances, if the diameter d_{211} is such that

$$d_{211}^{3/2} = d_{3111}^{3/2} + d_{3112}^{3/2} \tag{16}$$

then we may replace the two branches at Y_3, d_{3111} and d_{3112}, by one semi-infinite branch, which extends d_{211} to infinity; it is assumed here that R_m and R_i (hence K) are the same for all branches.

We can now proceed to do that same reduction at the branch point Y_2. If the diameter d_{11} is such that

$$d_{11}^{3/2} = d_{211}^{3/2} + d_{212}^{3/2} \tag{17}$$

then we may replace the two branches at Y_2, d_{211}, and d_{212}, by the branch d_{211} extended to infinity.

Proceeding in this way with the other branch points, we may also replace the part of the tree branching from d_{12} by the branch d_{12} extended to infinity. Finally, if the diameter d_0 is such that

$$d_0^{3/2} = d_{11}^{3/2} + d_{12}^{3/2} \tag{18}$$

then we may replace the two branches at Y_1 by the branch d_0 extended to infinity.

In conclusion, the Rall Theory asserts that if the dendritic branches satisfy the relations (16), (17), ..., (18), then the dendritic tree is equivalent to one semi-infinite core conductor of diameter do. This result holds for general dendritic trees provided

$$d_P^{3/2} = \Sigma d_D^{3/2}$$

when d_P is in any *parent* dendritic branch and d_D varies over its *daughter* dendritic branches.

The above analysis extends to the case where all the end-branches are of the same length L and all other branches are of length smaller than L. In this case, formula (15) is replaced by

$$G = Kd^{3/2} \tanh L .$$

References

1. Johnston, D., & Miao-Sin Wu, S. (2001). *Foundations of Cellular Neurophysiology*. Cambridge, MA: MIT Press.
2. Kandel, E.R., Schwartz, J.H, & Jessell, T.M. (1995). *Essentials of Neural Science Behavior*. New York: McGraw-Hill.
3. Nicholls, J.G., Martin, A.R., Wallace, B.G., & Fuchs, P.A. (2001). *From Neuron to Brain* (4th ed.). Sunderland, MA: Sinauer Associates Publishers.
4. Scott, A. (2002). *Neuroscience, A Mathematical Primer*. New York: Springer.

References [1] and [3] give a descriptive theory of neuroscience, and references [2] and [4] focus more on the modeling and the mathematical/computational aspects of neurons.

An Introduction to Dynamical Systems and Neuronal Dynamics

David Terman

Department of Mathemtaics, Ohio State University
231 West 18th Ave., 3210-1174 Columbus, USA
terman@math.ohio-state.edu

1 Introduction

A fundamental goal of neuroscience is to understand how the nervous system communicates and processes information. The basic structural unit of the nervous system is the individual neuron which conveys neuronal information through electrical and chemical signals. Patterns of neuronal signals underlie all activities of the brain. These activities include simple motor tasks such as walking and breathing and higher cognitive behaviors such as thinking, feeling and learning [18, 16].

Of course, neuronal systems can be extremely complicated. There are approximately 10^{12} neurons in the human brain. While most neurons consist of dendrites, a soma (or cell body), and an axon, there is an extraordinary diversity of distinct morphological and functional classes of neurons. Moreover, there are about 10^{15} synapses; these are where neurons communicate with one another. Hence, the number of synaptic connections made by a neuron can be very large; a mammalian motor neuron, for example, receives inputs from about 10^4 synapses.

An important goal of mathematical neuroscience is to develop and analyze mathematical models for neuronal activity patterns. The models are used to help understand how the activity patterns are generated and how the patterns change as parameters in the system are modulated. The models can also serve to interpret data, test hypotheses, and suggest new experiments. Since neuronal systems are typically so complicated, one must be careful to model the system at an appropriate level. The model must be complicated enough so that it includes those processes which are believed to play an important role in the generation of a particular activity pattern; however, it cannot be so complicated that it is impossible to analyze, either analytically or computationally.

A neuronal network's population rhythm results from interactions between three separate components: the intrinsic properties of individual neurons, the synaptic properties of coupling between neurons, and the architecture of coupling (i.e., which neurons communicate with each other). These components typically involve numerous parameters and multiple time scales. The synaptic coupling, for example, can be excitatory or inhibitory, and its possible turn on and turn off rates can vary widely. Neuronal systems may include several different types of cells as well as different types of coupling. An important and typically very challenging problem is to determine the role each component plays in shaping the emergent network behavior.

Models for the relevant neuronal networks often exhibit a rich structure of dynamic behavior. The behavior of even a single cell can be quite complicated. An individual cell may, for example, fire repetitive action potentials or bursts of action potentials that are separated by silent phases of near quiescent behavior [27, 15]. Examples of population rhythms include synchronized oscillations, in which every cell in the network fires at the same time and clustering, in which the entire population of cells breaks up into subpopulations or blocks; every cell within a single block fires synchronously and different blocks are desynchronized from each other [10, 31]. Of course, much more complicated population rhythms are possible. The activity may, for example, propagate through the network in a wave-like manner, or exhibit chaotic dynamics [29, 44, 42].

In this article, I will discuss models for neuronal systems and dynamical systems methods for analyzing these models. The discussion will focus primarily on models which include a small parameter and results in which geometric singular perturbation methods have been used to analyze the network behavior. I will not consider other types of models which are commonly used in the study of neural systems. The integrate and fire model of a single cell is one such example. A review of these types of models can be found in [20, 13].

An outline of the article is as follows. Chapters 2 and 3 present an informal introduction to the geometric theory of dynamical systems. I introduce the notions of phase space, local and global bifurcation theory, stability theory, oscillations, and geometric singular perturbation theory. All of these techniques are very important in the analysis of models for neuronal systems. Chapter 4 presents some of the basic biology used in modeling the neuronal systems. I will then discuss the explicit equations for the networks to be considered. Models for single cells are based on the Hodgkin-Huxley formalism [12] and the coupling between cells is meant to model chemical synapses. I will then consider models for single cells that exhibit bursting oscillations. There are, in fact, several different types of bursting oscillations, and there has been considerable effort in trying to classify the underlying mathematical mechanisms responsible for these oscillations [27, 15]. I then discuss the dynamics of small networks of neurons. Conditions will be given for when these networks exhibit either synchronous or desynchronous rhythms. I conclude by discussing an example of a larger network. This network was introduced as a model for activity patterns in the Basal Ganglia, a part of the brain involved in the generation of movements.

2 One Dimensional Equations

2.1 The Geometric Approach

This chapter and the next provide an informal introduction to the dynamical systems approach for studying nonlinear, ordinary differential equations. A more thorough presentation can be found in [36], for example. This approach associates a picture (the phase space) to each differential equation. Solutions, such as a resting state or oscillations, correspond to geometric objects, such as points or curves, in phase space. Since it is usually impossible to derive an explicit formula for the solution of a nonlinear equation, the phase space provides an extremely useful way for understanding qualitative features of solutions. In fact, even when it is possible to write down a solution in closed form, the geometric phase space approach is often a much easier way to analyze an equation. We illustrate this with the following example.

Consider the first order, nonlinear differential equation

$$\frac{dx}{dt} = x - x^3 \equiv f(x). \tag{1}$$

Note that it is possible to solve this equation in closed form by separating variables and then integrating. The resulting formula is so complicated, however, that it is difficult to interpret. Suppose, for example, we are given an initial condition, say $x(0) = \pi$, and we asked to determine the behavior of the solution $x(t)$ as $t \to \infty$. The answer to this question is not at all obvious by considering the solution formula.

The geometric approach provides a simple solution to this problem and is illustrated in Fig. 1. We think of $x(t)$ as the position of a particle moving along the x-axis at some time t. The differential equation gives us a formula

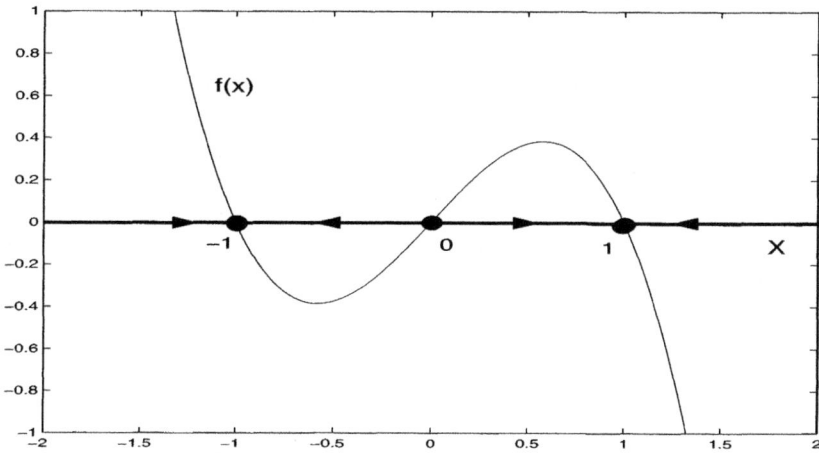

Fig. 1. The phase space for Equation (1).

for the velocity $x'(t)$ of the particle; namely, $x'(t) = f(x)$. Hence, if at time t, $f(x(t)) > 0$, then the position of the particle must increase, while if $f(x(t)) < 0$, then the position must be decrease.

Now consider the solution that begins at $x(0) = \pi$. Since $f(\pi) = \pi - \pi^3 < 0$, the solution initially decreases, moving to the left. It continues to move to the left and eventually approaches the fixed point at $x = 1$. A *fixed point* is a value of x where $f(x) = 0$.

This sort of analysis allows us to understand the behavior of every solution, no matter what its initial position. The differential equation tells us what the velocity of a particle is at each position x. This defines a vector field; each vector points either to the right or to the left depending on whether $f(x)$ is positive or negative (unless x is a fixed point). By following the position of a particle in the direction of the vector field, one can easily determine the behavior of the solution corresponding to that particle.

A fixed point is *stable* if every solution initially close to the fixed point remains close for all positive time. (Here we only give a very informal definition.) The fixed point is *unstable* if it is not stable. In this example, $x = -1$ and $x = 1$ are stable fixed points, while $x = 0$ is unstable.

This analysis carries over for *every* scalar differential equation of the form $x' = f(x)$, no matter how complicated the nonlinear function $f(x)$ is. Solutions can be thought of as particles moving along the real axis depending on the sign of the velocity $f(x)$. Every solution must either approach a fixed point as $t \to \pm\infty$ or become unbounded. It is not hard to realize that a fixed point x_0 is stable if $f'(x_0) < 0$ and is unstable if $f'(x_0) > 0$. If $f'(x_0) = 0$, then one must be careful since x_0 may be stable or unstable.

2.2 Bifurcations

Bifurcation theory is concerned with how solutions of a differential equation depend on a parameter. Imagine, for example, that an experimentalist is able to control the level of applied current injected into a neuron. As the level of applied current increases, the neuron may switch its behavior from a resting state to exhibiting sustained oscillations. Here, the level of applied current represents the bifurcation parameter. Bifurcation theory (together with a good model) can explain how the change in dynamics, from a resting state to oscillations, takes place. It can also be used to predict the value of injected current at which the neuron begins to exhibit oscillations. This may be a useful way to test the model.

There are only four major types of so-called local bifurcations and three of them can be explained using one-dimensional equations. We shall illustrate each of these with a simple example. The fourth major type of local bifurcation is the *Hopf bifurcation*. It describes how stable oscillations arise when a fixed point loses its stability. This requires at least a two dimensional system and is discussed in the next chapter.

Saddle-Node Bifurcation

The following example illustrates the *saddle-node bifurcation*:

$$x' = \lambda + x^2. \tag{2}$$

Here, λ is a fixed (bifurcation) parameter and may be any real number. We wish to solve this equation for a given value of λ and to understand how qualitative features of solutions change as the bifurcation parameter is varied.

Consider, for example, the fixed points of (2) for different values of the bifurcation parameter. Recall that fixed points are those values of x where the right hand side of (2) is zero. If $\lambda < 0$ then (2) has two fixed points; these are at $x = \pm\sqrt{-\lambda}$. If $\lambda = 0$ then there is only one fixed point, at $x = 0$, and if $\lambda > 0$ then there are no fixed points of (2).

To determine the stability of the fixed points, we let $f_\lambda(x) \equiv \lambda + x^2$ denote the right hand side of (2). A fixed point x_0 is stable if $f_\lambda'(x_0) < 0$. Here, differentiation is with respect to x. Since $f_\lambda'(x) = 2x$, it follows that the fixed point at $-\sqrt{-\lambda}$ is stable and the fixed point at $+\sqrt{-\lambda}$ is unstable.

A very useful way to visualize the bifurcation is shown in Fig 2 (left). This is an example of a *bifurcation diagram*. We plot the fixed points $x = \pm\sqrt{-\lambda}$ as functions of the bifurcation parameter. The upper half of the fixed point curve is drawn with a dashed line since these points correspond to unstable fixed points, and the lower half is drawn with a solid line since these points correspond to stable fixed points. The point $(\lambda, x) = (0, 0)$ is said to be a *bifurcation point*. At a bifurcation point there is a qualitative change in the nature of the fixed point set as the bifurcation parameter varies.

A basic feature of the saddle-node bifurcation is that as the bifurcation parameter changes, two fixed points, one stable and the other unstable, come together and annihilate each other. A closely related example is $x' = -\lambda + x^2$. There are no fixed points for $\lambda < 0$ and two for $\lambda > 0$. Hence, two fixed points are created as λ increases through the bifurcation point at $\lambda = 0$. This is also referred to as a saddle-node bifurcation.

Transcritical Bifurcation

A second type of bifurcation is the *transcritical bifurcation*. Consider the equation

$$x' = \lambda x - x^2. \tag{3}$$

Note that $x = 0$ is a fixed point for all values of λ; moreover, there is a second fixed point at $x = \lambda$.

To determine the stability of the fixed points, we let $f_\lambda(x) \equiv \lambda x - x^2$ denote the right hand side of (3). Since $f_\lambda'(x) = \lambda - 2x$, it follows that the fixed point at $x = 0$ is stable if $\lambda < 0$ and is unstable if $\lambda > 0$. The fixed point at $x = \lambda$ is stable if $\lambda > 0$ and is unstable if $\lambda < 0$.

The bifurcation diagram corresponding to this equation is shown in Fig. 2 (right). As before, we plot values of the fixed points versus the bifurcation

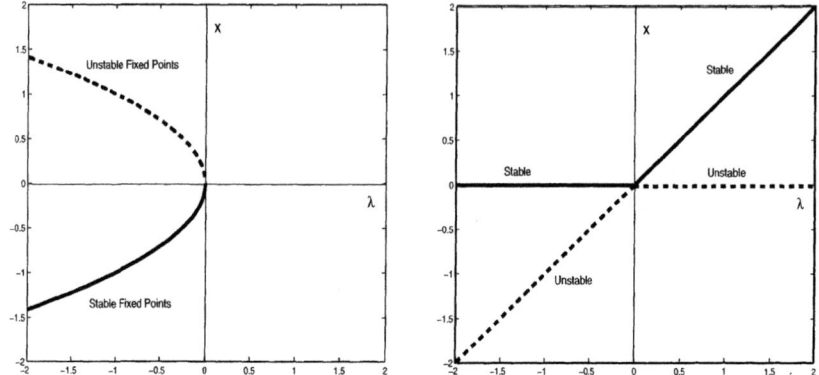

Fig. 2. The saddle-node bifurcation (left) and the transcritical bifurcation (right).

parameter λ. Solid curves represent stable fixed points, while dashed curves represent unstable fixed points. Note that there is an *exchange of stability* at the bifurcation point $(\lambda, x) = (0, 0)$ where the two curves cross.

Pitchfork Bifurcation

The third type of bifurcation is the *pitchfork bifurcation*. Consider the equation

$$x' = \lambda x - x^3. \tag{4}$$

If $\lambda \leq 0$, then there is one fixed point at $x = 0$. If $\lambda > 0$, then there are three fixed points. One is at $x = 0$ and the other two satisfy $x^2 = \lambda$.

In order to determine the stability of the fixed points, we let $f_\lambda(x) \equiv \lambda x - x^3$. Note that $f'_\lambda(x) = \lambda - 3x^2$. It follows that $x = 0$ is stable for $\lambda < 0$ and unstable for $\lambda > 0$. Moreover, if $\lambda > 0$ then both fixed points $x = \pm\sqrt{\lambda}$ are stable.

The bifurcation diagram corresponding to (4) is illustrated in Fig. 3 (left). There are actually two types of pitchfork bifurcations; (4) is an example of the *supercritical* case. An example of a *subcritical* pitchfork bifurcation is

$$x' = \lambda x + x^3. \tag{5}$$

The bifurcation diagram for this equation is shown in Fig 3 (right). Here, $x_0 = 0$ is a fixed point for all λ. It is stable for $\lambda < 0$ and unstable for $\lambda > 0$. If $\lambda < 0$, then there are two other fixed points; these are at $x_0 = \pm\sqrt{-\lambda}$. Both of these fixed points are unstable.

2.3 Bistability and Hysteresis

Our final example of a scalar ordinary differential equation is:

$$x' = \lambda + 3x - x^3. \tag{6}$$

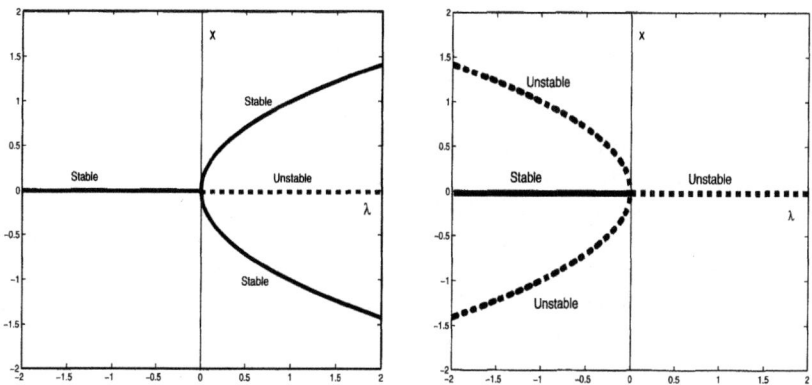

Fig. 3. A supercritical pitchfork bifurcation (left) and a subcritical pitchfork bifurcation (right).

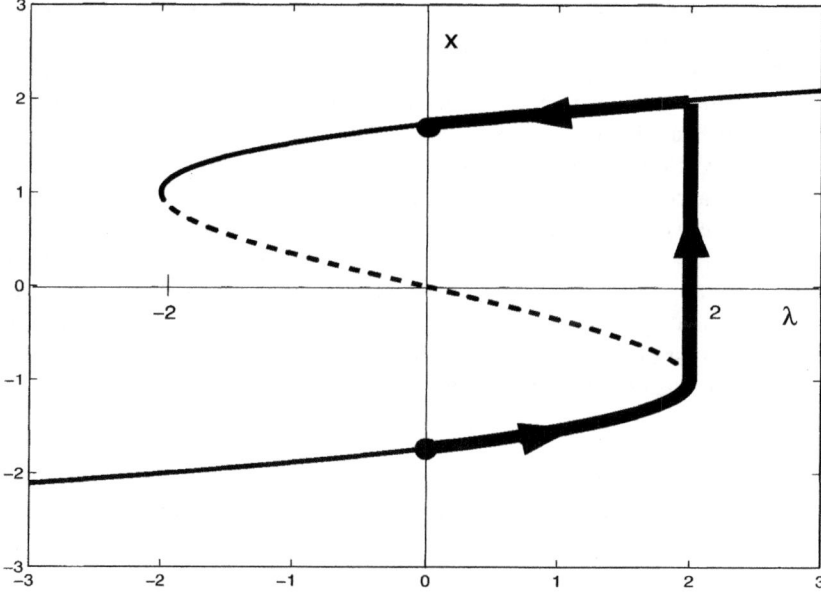

Fig. 4. Example of hysteresis.

The bifurcation diagram corresponding to (6) is shown in Fig. 4. The fixed points lie along the cubic $x^3 - 3x - \lambda = 0$. There are three fixed points for $|\lambda| < 2$ and one fixed point for $|\lambda| > 2$. We note that the upper and lower branches of the cubic correspond to stable fixed points, while the middle branch corresponds to unstable fixed points. Hence, if $|\lambda| < 2$ then there are two stable fixed points and (6) is said to be *bistable*.

There are two bifurcation points. These are at $(\lambda, x) = (-2, 1)$ and $(\lambda, x) = (2, -1)$ and both correspond to saddle-node bifurcations.

Suppose we slowly change the parameter λ, with initially $\lambda = 0$ and x at the stable fixed point $-\sqrt{3}$. As λ increases, (λ, x) remains close to the lower branch of stable fixed points. (See Fig. 4.) This continues until $\lambda = 2$ when (λ, x) crosses the saddle-node bifurcation point at $(\lambda, x) = (2, -1)$. The solution then approaches the stable fixed point along the upper branch. We now decrease λ to its initial value $\lambda = 0$. The solution remains on the upper branch. In particular, $x = \sqrt{3}$ when $\lambda = 0$. Note that while λ has returned to its initial value, the state variable x has not. This is an example of what is often called a *hysteresis phenomenon*.

3 Two Dimensional Systems

3.1 The Phase Plane

We have demonstrated that solutions of first order differential equations can be viewed as particles flowing in a one dimensional phase space. Remarkably, there is a similar geometric interpretation for *every* ordinary differential equation. One can always view solutions as particles flowing in some higher dimensional Euclidean (or phase) space. The dimension of the phase space is closely related to the order of the ode. Trajectories in higher dimensions can be very complicated, much more complicated than the one dimensional examples considered above. In one dimension, solutions (other than fixed points) must always flow monotonically to the left or to the right. In higher dimensions, there is a much wider range of possible dynamic behaviors. Here, we consider two dimensional systems, where many of the techniques used to study higher dimensional systems can be introduced.

A two dimensional system is one of the form

$$x' = f(x, y)$$
$$y' = g(x, y). \tag{7}$$

Here, f and g are given (smooth) functions; concrete examples are considered shortly. The phase space for this system is simply the $x - y$ plane; this is usually referred to as the *phase plane*. If $(x(t), y(t))$ is a solution of (7), then at each time t_0, $(x(t_0), y(t_0))$ defines a point in the phase plane. The point changes with time, so the entire solution $(x(t), y(t))$ traces out a curve, or trajectory, in the phase plane.

Of course, not every arbitrarily drawn curve in the phase plane corresponds to a solution of (7). What is special about solution curves is that the velocity vector at each point along the curve is given by the right hand side of (7). That is, the velocity vector of the solution curve $(x(t), y(t))$ at a point (x_0, y_0) is given by $(x'(t), y'(t)) = (f(x_0, y_0), g(x_0, y_0))$. This geometric property – that the vector $(f(x, y), g(x, y))$ always points in the direction that the solution is flowing – completely characterizes the solution curves.

3.2 An Example

Consider the system

$$x' = y - x^2 + x$$
$$y' = x - y. \tag{8}$$

We wish to determine the behavior of the solution that begins at some pre-scribed initial point $(x(0), y(0)) = (x_0, y_0)$. This will be done by analyzing the phase plane associated with the equations.

We begin the phase plane analysis by considering the vector field defined by the right hand side of (8). This is shown in Fig.5 where we have drawn the vector $(y - x^3 + x, x - y)$ at a number of points (x, y). Certainly, one cannot draw the vector field at every point. By considering enough points, one can get a sense of how the vector field behaves, however. A systematic way of doing this is as follows.

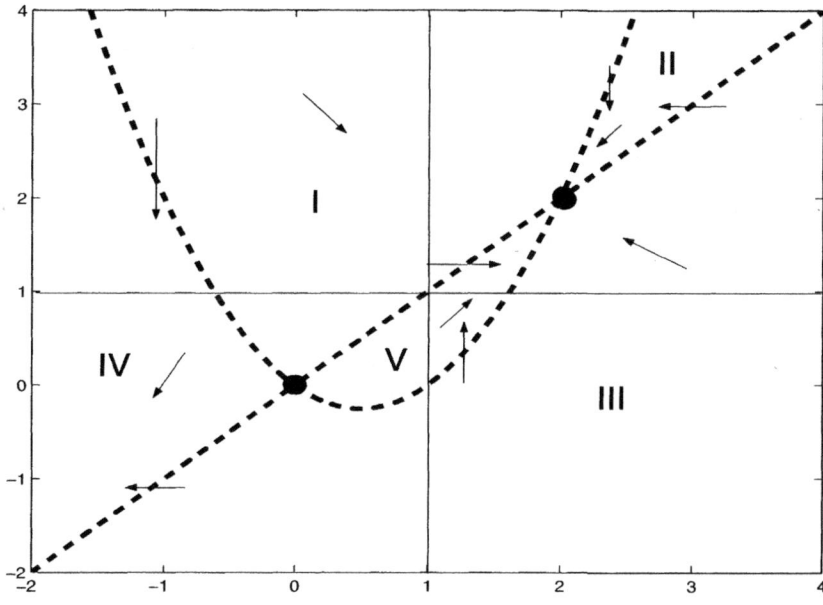

Fig. 5. The phase plane for equation (8).

The first step is to locate the fixed points. For a general system of the form (7), the fixed points are where both f and g vanish. For the example (8), there are two fixed points; these are at $(0,0)$ and $(2,2)$. Later we discuss later how one determines whether a fixed point is stable or unstable.

The next step is to draw the *nullclines*. The *x-nullcline* is where $x' = 0$; this is the curve $y = x^2 - x$. The *y-nullcline* is where $y' = 0$; this is the curve $y = x$. Note that fixed points are where the nullclines intersect. The nullclines divide the phase plane into five separate regions. All of the vectors within a given region point towards the same quadrant. For example, in the region labeled (I), $x' > 0$ and $y' < 0$. Hence, each vector points towards the fourth quadrant as shown. The vector field along the nullclines must be either horizontal or vertical. Along the x-nullcline, the vectors point either up or down, depending on the sign of y'. Along the y-nullcline, the vectors point either to the left or to the right, depending on the sign of x'.

It is now possible to predict the behavior of the solution to (8) with some prescribed initial condition (x_0, y_0). Suppose, for example, that (x_0, y_0) lies in the intersection of the first quadrant with region (I). Since the vector field points towards the fourth quadrant, the solution initially flows with $x(t)$ increasing and $y(t)$ decreasing. There are now three possibilities. The solution must either; (A) enter region II, (B) enter region V, or (C) remain in region I for all $t > 0$. It is not hard to see that in cases A or B, the solution must remain in region II or V, respectively. In each of these three cases, the solution must then approach the fixed point at $(2, 2)$ as $t \to \infty$.

We note that $(0, 0)$ is an unstable fixed point and $(2, 2)$ is stable. This is not hard to see by considering initial data close to these fixed points. For example, every solution that begins in region V must remain in region V and approach the fixed point at $(2, 2)$ as $t \to \infty$. Since one can choose points in region V that are arbitrarily close to $(0, 0)$, it follows that the origin must be unstable.

A more systematic way to determine the stability of the fixed points is to use the method of linearization. This method also allows us to understand the nature of solutions near the fixed point. Here we briefly describe how this important method works; this topic is discussed in more detail in any book on differential equations.

The basic idea of linearization is to replace the nonlinear system (7) by the linear one that best approximates the system near a given fixed point. One can then solve the linear system explicitly to determine the stability of the fixed point. If (x_0, y_0) is a fixed point of (7), then this linear system is:

$$x' = \frac{\partial f}{\partial x}(x_0, y_0)(x - x_0) + \frac{\partial f}{\partial y}(x_0, y_0)(y - y_0)$$

$$y' = \frac{\partial g}{\partial x}(x_0, y_0)(x - x_0) + \frac{\partial g}{\partial y}(x_0, y_0)(y - y_0). \tag{9}$$

Note that the right hand side of (9) represents the linear terms in the Taylor series of $f(x, y)$ and $g(x, y)$ about the fixed point. The stability of the fixed point is determined by the eigenvalues of the Jacobian matrix given by the partial derivatives of f and g with respect to x and y. If both eigenvalues have negative real part, then the fixed point is stable, while if at least one of the eigenvalues has positive real part, then the fixed point must be unstable.

By computing eigenvalues one easily shows that in the example given by (8), $(0, 0)$ is unstable and $(2, 2)$ is stable.

3.3 Oscillations

We say that a solution $(x(t), y(t))$ is *periodic* if $(x(0), y(0)) = (x(T), y(T))$ for some $T > 0$. A periodic solution corresponds to a closed curve or *limit cycle* in the phase plane. Periodic solutions can be either stable or unstable. Roughly speaking, a periodic solution is stable if solutions that begin close to the limit cycle remain close to the limit cycle for all $t > 0$. We do not give a precise definition here.

It is usually much more difficult to locate periodic solutions than it is to locate fixed points. Note that *every* ordinary differential equation can be written in the form $x' = f(x)$, $x \in R^n$ for some $n \geq 1$. A fixed point x_0 satisfies the equation $f(x_0) = 0$ and this last equation can usually be solved with straightforward numerical methods. We also note that a fixed point is a local object – it is simply one point in phase space. Oscillations or limit cycles are global objects; they correspond to an entire curve in phase space that retraces itself. This curve may be quite complicated.

One method for demonstrating the existence of a limit cycle for a two dimensional flow is the Poincare-Bendixson theorem [36]. This theorem does not apply for higher dimensional flows, so we shall not discuss it further. Three more general methods for locating limit cycles are the Hopf bifurcation theorem, global bifurcation theory and singular perturbation theory. These methods are discussed in the following sections.

3.4 Local Bifurcations

Recall that bifurcation theory is concerned with differential equations that depend on a parameter. We saw that one dimensional flows can exhibit saddle-node, transcritical and pitchfork bifurcations. These are all examples of local bifurcations; they describe how the structure of the flow changes near a fixed point as the bifurcation parameter changes. Each of these local bifurcations can arise in higher dimensional flows. In fact, there is only one major new type of bifurcation in dimensions greater than one. This is the so-called Hopf bifurcation. We begin this section by giving a necessary condition for the existence of a local bifurcation point. We then describe the Hopf bifurcation.

We consider systems of the form

$$x' = f(x, y, \lambda)$$
$$y' = g(x, y, \lambda). \tag{10}$$

It will be convenient to write this system using vector notation. Let

$$u = (x, y)^T \quad and \quad F(u, \lambda) = (f(x, y, \lambda), g(x, y, \lambda))^T.$$

Then (10) becomes

$$u' = F(u, \lambda). \tag{11}$$

We note that nothing described here depends on (10) being a two dimensional system. The following characterization of a local bifurcation point holds in arbitrary dimensions.

Suppose that u_0 is a fixed point of (10) for some value, say λ_0, of the bifurcation parameter. This simply means that $F(u_0, \lambda_0) = 0$. We will need to consider the Jacobian matrix J of F at u_0. We say that u_0 is a *hyperbolic fixed point* if J does not have any eigenvalues on the imaginary axis. An important result is that if u_0 is hyperbolic, then (u_0, λ_0) cannot be a bifurcation point. That is, a necessary condition for (u_0, λ_0) to be a bifurcation point is that the Jacobian matrix has purely imaginary eigenvalues. Of course, the converse statement may not be true.

We now describe the Hopf bifurcation using an example. Consider the system

$$x' = 3x - x^3 - y$$
$$y' = x - \lambda. \tag{12}$$

Note that there is only one fixed point for each value of the bifurcation parameter λ. This fixed point is at $(x, y) = (\lambda,\ 3\lambda - \lambda^3)$. It lies along the left or right branch of the cubic x-nullcline if $|\lambda| > 1$ and lies along the middle branch of this cubic if $|\lambda| < 1$.

We linearize (12) about the fixed point and compute the corresponding eigenvalues to find that the fixed point is stable for $|\lambda| > 1$ and unstable for $|\lambda| < 1$. When $|\lambda| = 1$, the fixed points are at the local maximum and local minimum of the cubic; in this case, the eigenvalues are $\pm i$. In particular, the fixed points are not hyperbolic and a bifurcation is possible when $\lambda = \pm 1$.

As λ increases past -1, the fixed point loses its stability. The eigenvalues are complex, so trajectories spiral towards the fixed point for $\lambda < -1$ and trajectories spiral away from the fixed point for $\lambda > -1$. (Here we are assuming that $|\lambda + 1|$ is not too large.) One can show (using a computer) that these unstable trajectories must approach a stable limit cycle. The amplitude of the limit cycle approaches zero as $\lambda \to -1$.

This is an example of a Hopf bifurcation. As the bifurcation parameter varies, a fixed point loses its stability as its corresponding eigenvalues cross the imaginary axis. The Hopf Bifurcation Theorem gives precise conditions for when this guarantees the existence of a branch of periodic orbits.

Note that (12) exhibits two Hopf bifurcations. The first is the one we have just discussed. It takes place when $\lambda = -1$ and the fixed point $(x_0, y_0) = (-1, -2)$ is at the local minimum of the cubic x-nullcline. The second Hopf bifurcation takes place when $\lambda = +1$ and the fixed point $(x_0, y_0) = (1, 2)$ is at the local maximum of the cubic x-nullcline. Figure 6 shows a bifurcation diagram corresponding to (12). Here we plot the maximum value of the x-variable along a solution as a function of the bifurcation parameter λ. The line $x = \lambda$ corresponds to fixed points. This is drawn as a bold, solid line for $|\lambda| > 1$ since these points correspond to stable fixed points, and as a dashed

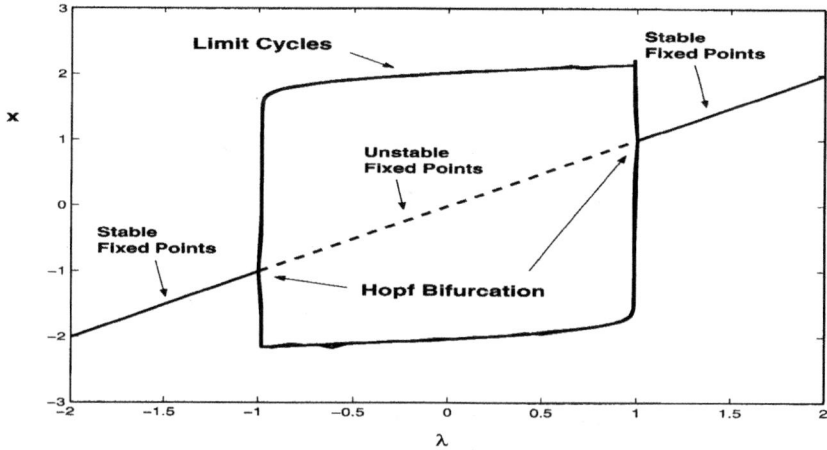

Fig. 6. The bifurcation diagram for (12). There are two Hopf bifurcation points.

line for $|\lambda| < 1$ since these points correspond to unstable fixed points. There is a curve corresponding to limit cycles that connects the bifurcation points at $(\lambda, x) = (-1, -1)$ and $(1, 1)$.

A Hopf bifurcation may be subcritical or supercritical. In the supercritical case, the limit cycles are stable and they exist for the same parameter values as the unstable fixed points (near the bifurcation point). In the subcritical case, the limit cycles are unstable and exist for those same parameter values as the stable fixed points.

3.5 Global Bifurcations

Hopf bifurcations are local phenomena; they describe the creation of limit cycles near a fixed point. As the bifurcation parameter approaches some critical value, the limit cycle approaches the fixed point and the amplitude of the limit cycle approaches zero. There are also global mechanisms by which oscillations can be created or destroyed. It is possible, for example, that the amplitude of oscillations remain bounded away from zero, but the frequency of oscillations approaches zero. This will be referred to as a *homoclinic* bifurcation (for reasons described below). It is also possible for two limit cycles, one stable and the other unstable, to approach and annihilate each other at some critical parameter value. This is referred to as a saddle-node bifurcation of limit cycles and resembles the saddle-node bifurcation of fixed points in which two fixed points come together and annihilate each other.

Fig. 7 illustrates a homoclinic bifurcation. For all values of the bifurcation parameter λ there are three fixed points; these are labeled as l, m, and u, and they are stable, a saddle and unstable, respectively. When $\lambda = \lambda_0$ (shown in the middle panel), there is a homoclinic orbit labeled as $\gamma_h(t)$. This orbit lies in both the stable and unstable manifolds of the fixed point m. If $\gamma < \gamma_0$

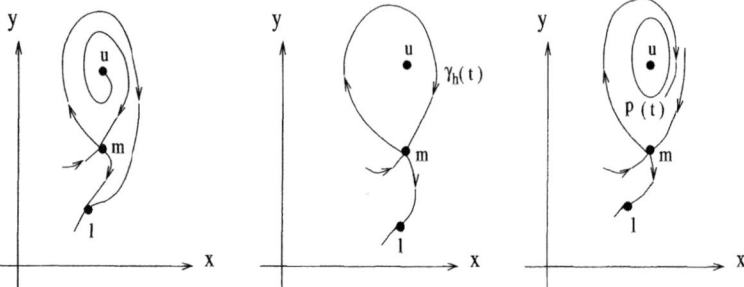

Fig. 7. A homoclinic bifurcation.

(shown in the left panel) then there is no periodic orbit, while if $\gamma > \gamma_0$, then there is a stable limit cycle, labeled as $p(t)$. Note that if $\gamma < \gamma_0$, then the stable manifold of m lies inside of the unstable manifold, while the opposite holds if $\gamma > \gamma_0$.

Note that solutions move very slowly as they pass near an unstable fixed point. It follows that the period of the periodic solution, for $\lambda > \lambda_0$, must become arbitrarily large as λ approaches λ_0. We also note that the nature of the three fixed points do not change as λ is varied. Hence, there is no local bifurcation. The homoclinic orbit is a global object, and the limit cycle disappears via a global bifurcation.

3.6 Geometric Singular Perturbation Theory

Models for neuronal systems often involve variables that evolve on very different time scales. We shall see many examples of such systems in the next chapter. The existence of different time scales naturally leads to models that contain small parameters. Geometric singular perturbation theory provides a powerful technique for analyzing these models. The theory gives a systematic way to reduce systems with small parameters to lower dimensional reduced systems that are more easily analyzed. Here we illustrate how this method works with a simple example. The method will be used extensively in the next chapter to study more complicated models arising from neuronal systems.

Consider a general two-dimensional system of the form

$$
\begin{aligned}
v' &= f(v, w) \\
w' &= \epsilon g(v, w).
\end{aligned}
\tag{13}
$$

Here, $\epsilon > 0$ is the small, singular perturbation parameter. We assume that the v−nullcline is a cubic-shaped curve and the w−nullcline is a monotone increasing curve that intersects the cubic at a single fixed point, denoted by p_0, that lies along the middle branch of the cubic nullcline. We also need to assume that $v' > 0\ (< 0)$ below (above) the cubic $v-$nullcline and $w' > 0\ (< 0)$ below (above) the $w-$nullcline.

One can prove, using the Poincare-Bendixson theorem, that (13) has a limit cycle for all ϵ sufficiently small. Moreover, the limit cycle approaches a singular limit cycle as shown in Fig. 8 as $\epsilon \to 0$. The singular limit cycle consists of four pieces. One of these pieces lies along the left branch of the cubic nullcline. We shall see in the next chapter that this corresponds to the silent phase of an action potential. Another piece of the singular solution lies along the right branch of the cubic nullcline; this corresponds to the active phase of the action potential. The other two pieces are horizontal curves in the phase plane and they connect the left and right branches. The "jump-up" to the active phase occurs at the left knee of the cubic and the "jump-down" occurs at the right knee of the cubic.

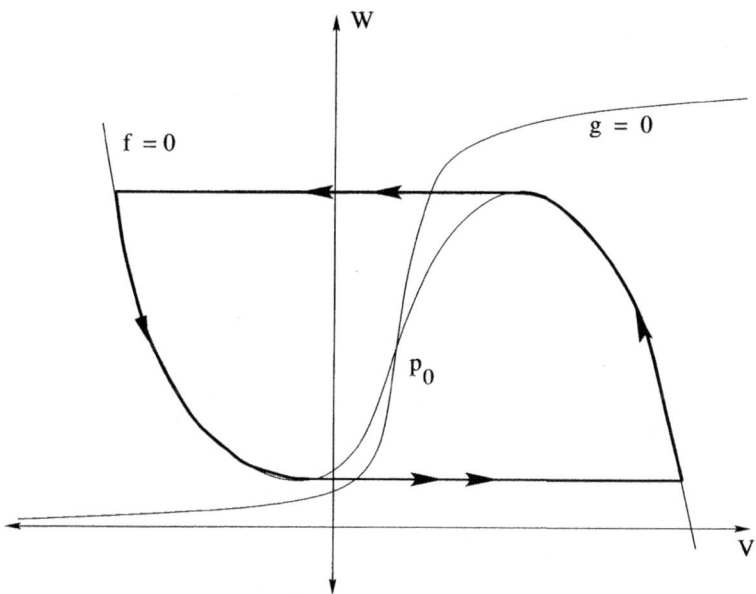

Fig. 8. Nullclines and singular periodic orbit for an oscillatory relaxation oscillator.

We refer to (13) as a *singular* perturbation problem because the structure of solutions of (13) with $\epsilon > 0$ is very different than the structure of solutions of (13) with $\epsilon = 0$. If we set $\epsilon = 0$, then (13) reduces to

$$v' = f(v, w)$$
$$w' = 0. \tag{14}$$

Note that w is constant along every solution of (14); that is, every trajectory is horizontal. The fixed point set is the entire cubic-shaped curve $\{f = 0\}$.

This is very different from (13) with $\epsilon > 0$ in which there is only one fixed point and w is not constant along all other solutions.

The reduced system (14) does, however, give a good approximation of solutions away from the cubic v−nullcline. In particular, it determines the evolution of the jump-up and jump-down portions of the singular solution. In order to determine the behavior of solutions near the cubic nullcline – that is, during the silent and active phases – we introduce the slow time scale $\tau = \epsilon t$ and then set $\epsilon = 0$ in the resulting system. The leads to the system

$$0 = f(v, w)$$
$$\dot{w} = g(v, w). \tag{15}$$

The first equation in (15) implies that the solution of (15) lies along the cubic nullcline. The second equation in (15) determines the evolution of the solution along this nullcline. Note that if we the write the left and right branches of the cubic nullcline as $v = H_L(w)$ and $v = H_R(w)$, respectively, then we can write the second equation in (15) as

$$\dot{w} = g(H_\alpha(w), w) \equiv G_\alpha(w) \tag{16}$$

where $\alpha = L$ or R.

Note that each piece of the singular solution is determined by a single, scalar differential equation. The silent and active phases correspond to solutions of (16). This equation will be referred to as the *slow equation*. The jump-up and jump-down correspond to solutions of the first equation in (14); this is referred to as the *fast equation*.

The analysis described here will be used to study more complicated, higher dimensional systems that arise in models for neuronal systems. Using the existence of small parameters, we will construct singular periodic solutions. Each piece of the singular solution will satisfy a reduced system of equations. We note that the order of the slow equations will be equal to the number of slow variables. In particular, if a given model has one variable that evolves at a much slower time-scale than the other variables, then the order of the slow equations will be just one. Hence, we will reduce the analysis of a complicated, high dimensional system to a single, scalar equation.

4 Single Neurons

In this chapter we discuss models for a single neuron. We begin by quickly reviewing the basic biology that forms the basis of the mathematical models. We then present the Hodgkin-Huxley equations. This is certainly the most important model in computational neuroscience. It was originally introduced as a model for the generation of action potentials in the giant axon of a squid and forms the basis of numerous other models of electrical activity in other neurons. We then present a simpler, reduced model. This will be very useful in our analysis of networks of neuronal activity. We then review geometric analysis of bursting oscillations.

4.1 Some Biology

The neuron is the basic information processing unit in the nervous system. Most neurons consist of a cell body (or soma) and a number of processes that project from the cell body; these are the dendrites and the axon. The dendrites spread out from the cell body in a tree-like manner. They collect incoming signals from other neurons or sensory receptors. Impulses are conducted away from the soma along the axon. Axons may be very short, while others may be very long (up to more than one meter). Many axons develop side branches called axon collaterals that help bring information to several parts of the nervous system simultaneously.

The neuron is surrounded by a cell membrane that maintains a stable resting potential between the outside and the inside of the cell. In response to a stimulus, the membrane potential may undergo a series of rapid changes, called an action potential. This results in the generation of a nerve impulse. In order to form a nerve impulse, the initial stimulus must be above some threshold amount. Properties of the nerve impulse, including its shape and propagation velocity, are often independent of the initial (superthreshold) stimulus.

The resting potential is created because there is an imbalance in the concentrations of certain ions between the inside and the outside of the cell. The intracellular concentrations of sodium and calcium ions are lower than in the extracellular space, while the extracellular potassium concentration is lower than inside the cell. In its resting state, the cell membrane is permeable to potassium; however it is virtually impermeable to sodium and calcium. The resting potential is about -70 mV and results, to a large extent, from the selective permeability of the membrane and the imbalance in concentration of potassium ions.

Following a stimulus, there is a sudden change in the permeability of the membrane to sodium (or calcium) ions. There are channels in the membrane selective to sodium. An action potential is generated when these channels open and sodium ions rush into the cell interior, causing a rapid rise in the membrane potential. After some delay, the sodium channels close and an increased number of potassium channels open. Potassium ions then flow to the outside of the cell and this brings the membrane towards the resting state. The membrane potential actually overshoots the resting potential. There is a recovery period in which sodium and potassium pumps move sodium and potassium ions, respectively, out of and into the cell until the resting membrane potential is achieved. During the initial phase of this recovery period, it is impossible to generate another action potential. This is referred to as the absolute refractory period.

4.2 The Hodgkin-Huxley Equations

The Hodgkin-Huxley equations were published in 1952 and describe the generation of action potentials in the squid giant axon [12]. The principles underlying the derivation of these equations form the basis of modeling other cells throughout the nervous system. The Hodgkin-Huxley model consists of four differential equations. One of these is a partial differential equation that describes the evolution of the membrane potential. The other three equations are ordinary differential equations that are related to properties of the ionic channels. The Hodgkin-Huxley equations can be written as:

$$C_M \frac{\partial v}{\partial t} = D_M \frac{\partial^2 v}{\partial x^2} - g_{Na} m^3 h(v - v_{Na}) - g_K n^4 (v - v_K) - g_L(v - v_L)$$

$$\frac{\partial m}{\partial t} = (m_\infty(v) - m)/\tau_m(v)$$

$$\frac{\partial n}{\partial t} = (n_\infty(v) - n)/\tau_n(v) \tag{17}$$

$$\frac{\partial h}{\partial t} = (h_\infty(v) - h)/\tau_h(v)$$

Here, $v(x,t)$ represents the membrane potential and each term in the first equation represents a separate current. Since the cell membrane separates charge it can be viewed as a capacitor and $C_M \frac{\partial v}{\partial t}$ is the capacitive current. The term $D_M \frac{\partial^2 v}{\partial x^2}$ represents longitudinal current along the axon and the remaining terms are ionic currents. The sodium current is $I_{Na} \equiv g_{Na} m^3 h(v - v_{Na})$. It is modeled using Ohm's law in which $g_{Na} m^3 h$ is the conductance and $(v - v_{Na})$ is a driving potential. The maximal sodium conductance is g_{Na} and $m^3 h$ can be thought of as the probability that a sodium channel is open. This will be discussed in more detail shortly. The constant v_{Na} is called the *sodium reversal potential*. This is the value of the membrane potential when the sodium concentration, which produces an inward flux of sodium through the sodium channel, is balanced by the electrical potential gradient tending to move sodium ions in the channel in the opposite direction. In a similar manner, $I_K \equiv g_K n^4(v - v_K)$ is the potassium current with g_K being the maximal potassium conductance and n^4 is the probability that a potassium channel is open. Finally, $I_L \equiv g_L(v - v_L)$ is usually referred to as a leak conductance; it is due to the effects of other, less important, ions including chloride.

Note that the gating variables m, h, and n satisfy differential equations that depend on the membrane potential v. Hence, the probability that a sodium or potassium channel is open or closed depends on the membrane potential. The sodium channel depends on two variables, namely m and h. We can think of the sodium channel as possessing two gates; both gates must be open in order for sodium to flow into the cell. We will not describe the voltage dependent steady-state functions $n_\infty(v), h_\infty(v), n_\infty(v)$ or the time constants $\tau_m(v), \tau_h(v), \tau_n(v)$ here. A detailed description can be found in [16, 19].

What distinguishes one cell from another are the types of ionic currents responsible for the action potential and what factors determine whether the channels corresponding to these ions are open or closed. In what follows, we simplify the presentation by ignoring the spatial dependence of the membrane potential; hence, we only consider systems of ordinary differential equations. Each cell then satisfies a system of the general form

$$C_M v' = -I_{ion}(v, w_1, w_2, ..., w_n) + I \tag{18}$$
$$w_i' = \epsilon[w_{i,\infty}(v) - w_i]/\tau_i(v).$$

Here $v(t)$ denotes the membrane potential, I_{ion} is the sum of v- and t-dependent currents through the various ionic channel types and I represents external applied current. Each variable $w_i(t)$ describes the state of channels of a given type. Each current I_j is assumed to be ohmic and can be expressed as $I_j = \hat{g}_j \sigma_j(v, w_1, ..., w_n)(v - v_j)$ where $\hat{g}_j > 0$, the v_j are constants and the function σ_j represents the fraction of j-channels that are open.

4.3 Reduced Models

The dynamics of even one single neuron can be quite complicated. Examples of such complex behavior are given in the next subsection when we discuss bursting oscillations. We are primarily interested in developing techniques to study networks consisting of possibly a large number of coupled neurons. Clearly the analysis of networks may be extremely challenging if each single element exhibits complicated dynamics. For this reason, we often consider simpler, reduced models for single neurons. The insights we gain from analyzing the reduced models are often extremely useful in studying the behavior of more complicated biophysical models.

An example of a reduced model are the Morris-Lecar equations [24]:

$$v' = -g_L(v - v_L) - g_K w(v - v_K) - g_{Ca} m_\infty(v)(v - v_{Ca}) + I$$
$$w' = \epsilon(w_\infty(v) - w)\cosh((v + .1)/.3) \tag{19}$$

The parameters and nonlinear functions in (19) are given by $v_L = -.1$, $g_L = .5$, $g_K = 2$, $v_K = -.7$, $g_{Ca} = 1$, $v_{Ca} = 1$, $\epsilon = .1$, $m_\infty(v) = .5(1 + \tanh((v - .01)/.145))$, $w_\infty(v) = .5(1 + \tanh((v + .1)/.15))$.

These equations can be viewed as a simple model for a neuron in which there are potassium and calcium currents along with the leak current. There is also an applied current represented by the constant I. Here, v represents the membrane potential, rescaled between -1 and 1, and w is activation of the potassium current. Note that the activation of the calcium current is instantaneous. The small, positive parameter ϵ is introduced to emphasize that w evolves on a slow time scale. We treat ϵ as a singular perturbation parameter in the analysis.

Note that this simple model has no spatial dependence; that is, we are viewing the neuron as a single point. This will certainly simplify our analysis of time-dependent oscillatory behavior. In Section 4.5, we add spatial dependence to the model and consider the generation of propagating wave activity.

If $I = 0$, then every solution approaches a stable fixed point. This corresponds to a neuron in its resting state. However, if $I = .4$, then solutions of (19) quickly approach a stable limit cycle. The periodic solution alternates between an active phase and a silent phase of near resting behavior. Moreover, there are sharp transitions between the silent and active phases.

These solutions can be easily analyzed using the phase space methods described in Section 3.6. The v-nullcline is a cubic-shaped curve, while the w-nullcline is a monotone increasing function that intersects the v-nullcline at precisely one point; hence, there exists precisely one fixed point of (19), denoted by p_0. If $I = 0$, then p_0 lies along the left branch of the cubic v-nullcline and one can easily show that p_0 is asymptotically stable. If, on the other hand, $I = .4$, then p_0 lies along the middle branch of the cubic nullcline and p_0 is unstable. There must then exist a stable limit cycle for all ϵ sufficiently small; moreover, the limit cycle approaches a singular limit cycle as shown in Fig. 8 as $\epsilon \to 0$.

The phase plane analysis does not depend on the precise details of the nonlinear functions and other parameters in (19). We will consider more general two dimensional systems of the form

$$v' = f(v, w) + I$$
$$w' = \epsilon g(v, w) \tag{20}$$

where the v-nullcline and the w-nullcline satisfy the assumptions described in Section 3.6.

, We note that, with some assumptions, one can systematically reduce a four dimensional Hodgkin-Huxley-like model (without spatial dependence) to a two dimensional Morris-Lecar-like model. One first observes that some of the channel-state variables evolve much faster than others. In particular, the sodium activation variable m evolves on a much faster time-scale than the sodium inactivation variable h and the potassium activation variable n. We therefore assume assume that m activates instantaneously; that is, we set $m = m_\infty(v)$. This reduces the Hodgkin-Huxley equations to just three equations for v, h, and n. To reduce the equations further, we observe that along solutions $h \approx \lambda(1-n)$ for some constant λ. Assuming this to be true, we can eliminate the equation for h and we now have a two-dimensional system similar to the Morris-Lecar equations.

The reduced two-dimensional models given by (19), or more generally (20), exhibit many properties of real neurons. For example, (20) exhibits a refractory period: immediately after an action potential it is difficult to generate

another one. This is because when the trajectory in phase space corresponding to the action potential returns to the silent phase, it lies along the left
branch of the cubic nullcline with an elevated value of the recovery variable
n. Here, the trajectory is further away from the threshold, corresponding to
the middle branch.

Note also that when there is no applied current (that is, $I = 0$), (20)
exhibits a stable fixed point; this corresponds to the resting state of a neuron.
If I is sufficiently large, then (20) exhibits sustained oscillations. The simple
model also displays *excitability*. That is, a small stimulus will not generate
an action potential. In this case the solution returns quickly to rest. In order
to produce an action potential, the initial stimulus must be larger than some
threshold. Note that the threshold corresponds to the position of the middle
branch of the cubic nullcline.

Finally, we now discuss how the geometric singular perturbation approach
can be used to understand the response of a neuron to injected current. Consider the system

$$v' = f(v, w) + I(t)$$
$$w' = \epsilon g(v, w) \tag{21}$$

where f and g are as in (20) and $I(t)$ represents the injected current. We
assume that when $I(t) = 0$ the system is excitable; that is, the v- and w-
nullclines intersect along the left branch of the cubic. This is a globally stable
fixed point and the model neuron is in its resting state. We further assume
that there exists I_0 and $T_{on} < T_{off}$ such that

$$I(t) = I_0 \quad \text{if} \quad T_{on} < t < T_{off} \quad \text{and} \quad I(t) = 0 \quad \text{otherwise.}$$

We will consider two cases: either $I_0 > 0$, in which case the injected current
is said to be *depolarizing*, or $I_0 < 0$ and the injected current is *hyperpolarizing*.
Fig. 9 illustrates the neuron's response when (top) $I_0 = .1$ and (bottom) $I_0 =
-.1$. In the depolarizing case, the neuron fires an action potential immediately
after the injected current is turned on. The cell then returns to rest. In the
hyperpolarizing case, the neuron's membrane potential approaches a more
negative steady state until the current is turned off, at which time the neuron
fires a single action potential. This last response is often called *post-inhibitory
rebound* [9].

The geometric approach is very useful in understanding these responses.
As before, we construct singular solutions in which ϵ is formally set equal to
zero. See Fig. 10. The singular solutions lie on the left or right branch of some
cubic-shaped nullcline during the silent and active phases. The cubics depend
on the values of $I(t)$. We denote the cubic corresponding to $I = 0$ as C and
the cubic corresponding to I_0 as C_0. Note that if $I_0 > 0$, then C_0 lies 'above'
C, while if $I_0 < 0$, then C_0 lies 'below' C. This is because of our assumption
that $f < 0$ (> 0) above (below) the v-nullcline.

Consider the depolarizing case $I_0 > 0$. This is illustrated in Fig. 10 (left).
For $t < T_{on}$, the solution lies at the fixed point p_0 along the left branch of C.

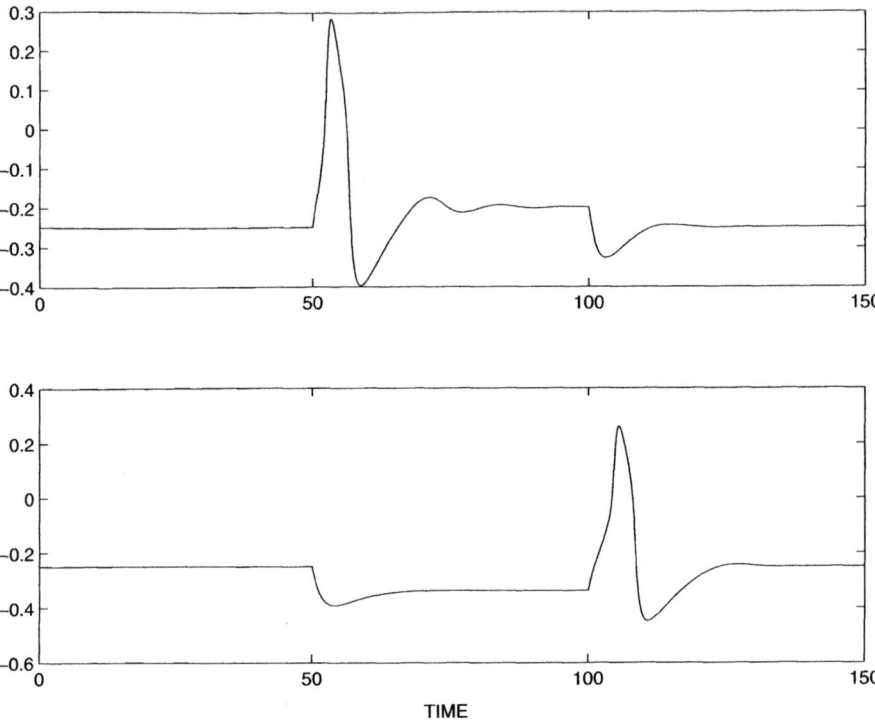

Fig. 9. Response of a model neuron to applied current. Current is applied at time $t = 50$ and turned off at $t = 100$. In the top figure, the current is depolarizing ($I_0 = .1$), while in the bottom figure the current is hyperpolarizing ($I_0 = -.1$) and the neuron exhibits post-inhibitory rebound.

When $t = T_{on}$, $I(t)$ jumps to I_0 and the cell's cubic switches to C_0. If the left knee of C_0 lies above p_0 then the cell jumps up to the right branch of C_0; this corresponds to the firing of an action potential. If the w-nullcline intersects C_0 along its left branch, then the cell will approach the stable fixed point along the left branch of C_0 until the input is turned off. It is possible that the w-nullcline intersects C_0 along its middle branch. If this is the case then the cell oscillates, continuing to fire action potentials; until $t = T_{off}$ when the input is turned off. Note that in order for the cell to fire an action potential, the injected current must be sufficiently strong. I_0 must be large enough so that the p_0 lies below the left knee of C_0.

We next consider the hyperpolarizing case $I_0 < 0$, shown in Fig. 10 (right). Then C_0 lies below C and the w-nullcline intersects C_0 at a point denoted by p_1: When $t = T_{on}$, the solution jumps to the left branch of C_0 and then evolves along this branch approaching p_1 for $T_{on} < t < T_{off}$. When $t = T_{off}$, I switches back to 0 and the cell now seeks the left or right branch of C. If, at this time, the cell lies below the left knee of C, then the cell will jump up

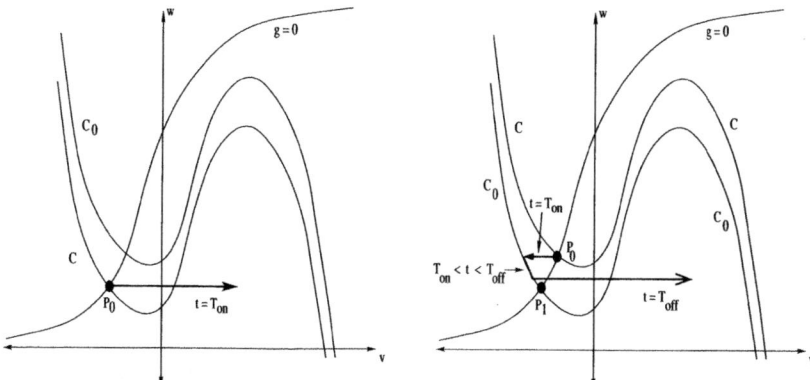

Fig. 10. Phase space representation of the response of a model neuron to applied current. Current is applied at time $t = T_{on}$ and turned off at $t = T_{off}$. (Left) Depolarizing current. The cell jumps up as soon as the current is turned on. (Right) Hyperplorizing current. The cell jumps to the left branch of C_0 when the current is turned on and jumps up to the active phase due to post-inhibitory rebound when the current is turned off.

to the active phase giving rise to post-inhibitory rebound. In order to have post-inhibitory rebound, the hyperpolarizing input must be sufficiently large and last sufficiently long. I_0 must be sufficiently negative so that p_1 lies below the left knee of C. Moreover, $T_{off} - T_{on}$ must be sufficiently large so that the cell has enough time to evolve along the left branch of C_0 so that it lies below the left knee of C when the input is turned off.

4.4 Bursting Oscillations

Certain neurons and other excitable cells exhibit bursting oscillations; this behavior is characterized by a silent phase of near steady state resting behavior alternating with an active phase of rapid, spike-like oscillations, as shown in Fig. 11. Examples of biological systems which display bursting oscillations include the Aplysia R-15 neuron, insulin secreting pancreatic beta cells, and neurons in the hippocampus, cortex and thalamus. For references, see [46], for example.

Fig. 11 shows three types of bursting oscillations. Fig. 11 (top) displays an example of *square-wave* bursting. This is characterized by abrupt periodic switching between the quiescent, or silent, phase and the active phase of repetitive firing. Note that the frequency of spikes decreases at the end of the active phase. Fig. 11 (middle) illustrates *elliptic* bursting. Small amplitude oscillations occur during the silent phase and the amplitude of spikes gradually waxes and wanes. Finally, Fig. 11 (bottom) displays *parabolic* bursting. The spike rate first increases and then decreases in a parabolic manner.

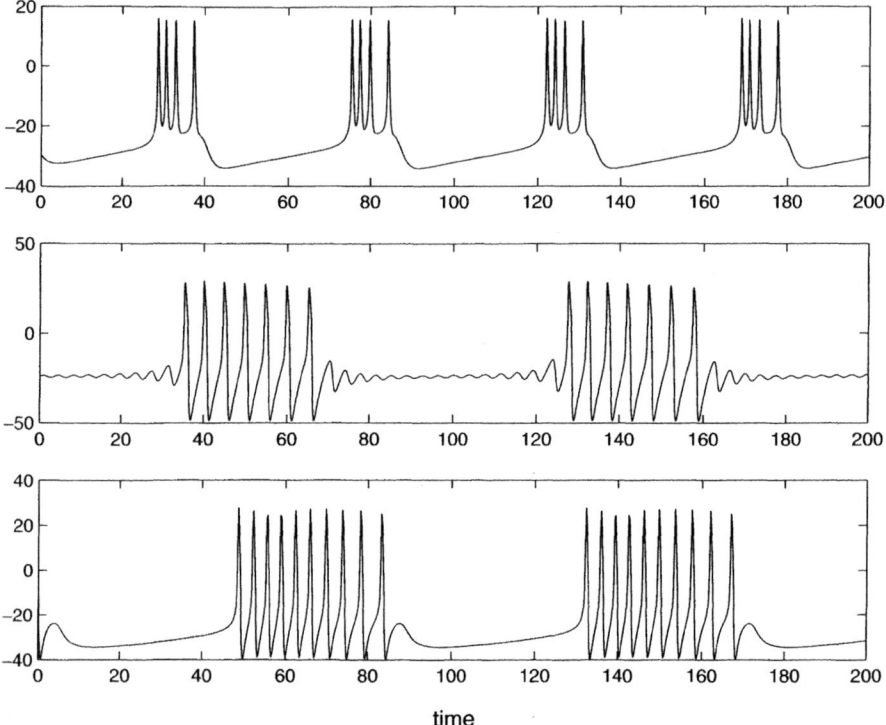

Fig. 11. Classes of bursting oscillations. (Top) Square-wave bursting. (Middle) Elliptic bursting. (Bottom) Parabolic bursting.

The mathematical mechanisms responsible for each class of bursting oscillation are described in terms of geometric properties of the corresponding phase space dynamics. Here we describe the mathematical mechanisms responsible for square-wave bursting. We only consider the simplest, lowest dimensional, models which generate square-wave bursting. Geometric analysis of other bursting types can be found in [27].

Consider a three-dimensional version of (13) of the form

$$
\begin{aligned}
v' &= f(v, w, y) \\
w' &= g(v, w, y) \\
y' &= \epsilon h(v, w, y, \lambda)
\end{aligned}
\tag{22}
$$

Here, $\epsilon > 0$ is a small, singular perturbation parameter and λ is some other fixed parameter. If we set $\epsilon = 0$, then y is constant along solutions and we can think of y as a bifurcation parameter in the fast system (FS) consisting of the first two equations in (22). The primary assumptions on (22) concern the bifurcation structure of the fast subsystem. This structure appears in Fig. 12, which also shows the projection of the square-wave bursting solution

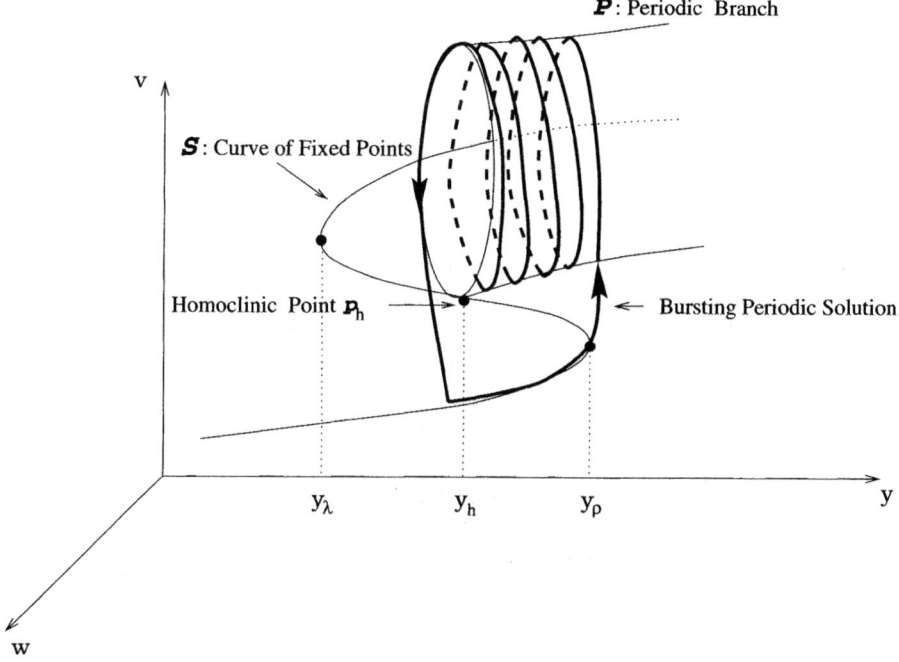

Fig. 12. Bifurcation structure for square-wave bursting. The bold curve is the projection of a square-wave bursting solution.

onto the corresponding bifurcation diagram. The set of fixed points of (FS) is assumed to be a Z-shaped curve in the (v, w, y) phase space. We denote this curve by S; only a portion of this Z-shaped curve is shown in Fig. 12. The fixed points along the lower branch of S are stable solutions of (FS), while the fixed points on the middle branch of S are saddles. Fixed points along the upper branch of S may be stable or unstable. We also assume that there exists a one-parameter family of periodic solutions of (FS), denoted by P. These limit cycles originate at a Hopf bifurcation along the upper branch of S and terminate along a solution of (FS) that is homoclinic to one of the fixed points on the middle branch of S.

Assumptions are also needed about the slow dynamics. We assume that the y-nullsurface $\{h = 0\}$ defines a two-dimensional manifold that intersects S at a single point. This point lies on the middle branch of S between the homoclinic point and the right knee of S. Finally, $h > 0$ above $\{h = 0\}$ and $h < 0$ below $\{h = 0\}$.

We now give a heuristic explanation for why this system generates a square-wave bursting solution. Suppose that $\epsilon > 0$ is small and consider a solution that begins close to the lower branch. Because this branch consists of stable fixed points of (FS), the trajectory quickly approaches a small neighborhood of the lower branch. The trajectory tracks rightward along the lower branch according to the slow dynamics, until it passes the right knee. This portion of the solution corresponds to the silent phase. Once past the right knee, the trajectory is attracted to near P, the branch of periodic solutions of (FS). This generates the fast repetitive spikes of the bursting solutions. The trajectory passes near P, with decreasing y, until it reaches a neighborhood of the homoclinic orbit of (FS). Once it passes the homoclinic orbit, the fast dynamics eventually forces the trajectory back to near the lower branch of S and this completes one cycle of the bursting solution.

This description is formal. It is not at all clear that if the system (22) satisfies the above assumptions, then, for all ϵ sufficiently small, there exists a unique periodic solution corresponding to a bursting oscillation. In fact, it is shown in [38] that such a result cannot be true, in general. However, in [22], it is proved that the bursting solution will be uniquely determined for all ϵ sufficiently small, except for those ϵ that lie in a certain very small set.

Remark 4.1 A crucial ingredient for square-wave bursting is bistability. This allows for a hysteresis loop between a lower branch of stable fixed points and an upper branch of stable limit cycles. It is also very important that the slow nullsurface $\{h = 0\}$ lies between these two branches. If this last condition is not satisfied, then the system may exhibit other types of solutions. For example, suppose that $\{h = 0\}$ intersects the lower branch of S. This point of intersection will then be a globally stable fixed point of (22). If, on the other hand, $\{h = 0\}$ intersects S along its middle branch above the homoclinic point, then (22) may give rise to a stable limit cycle which always remains in the active phase near P. This type of solution is referred to as *continuous spiking*. Rigorous results concerning the existence of continuous spiking are presented in [39].

Remark 4.2 Square-wave bursting arises in models for electrical activity in pancreatic β-cells. It is believed that this activity plays an important role in the release of insulin from the cells. The first mathematical model for this bursting was due to Chay and Keizer [5]. There have been numerous related models, based on experimental data, since then. A review of these models, along with a detailed description of the more biological issues, is given in [32]. Square wave bursting also arises in recent models for respiratory CPG neurons [3] and models for pattern generation based on synaptic depression [37].

Remark 4.3 Very complicated (global) bifurcations can take place as the parameters ϵ or λ are varied in (22). The singular perturbation parameter ϵ controls the rate at which a bursting trajectory passes through the silent and active phases. In particular, the number of spikes per burst is $O(1/\epsilon)$ and becomes unbounded as $\epsilon \to 0$. It is demonstrated in [38] that Smale horseshoe chaotic dynamics can arise during the transition of adding a spike.

Perhaps even more interesting is the bifurcation structure of (22) as λ is varied. In the β-cell models, λ is related to the glucose concentration. As the glucose level gradually increases, the cells exhibit resting behavior, then bursting oscillations, and then continuous spiking. This is consistent with behavior exhibited by the model. As λ increases, the y-nullsurface $\{h = 0\}$ intersects the lower branch of S, then the middle branch of S below the homoclinic point, and then the middle branch of S above the homoclinic point. Numerical studies [6] and rigorous analysis [39] have shown that as λ varies between the bursting and continuous spiking regimes, the bifurcation structure of solutions must be very complicated. A Poincaré return map defined by the flow from a section transverse to the homoclinic orbit of (FS) will exhibit Smale-horseshoe dynamics for a robust range of parameter values. This leads to solutions in which the number of spikes per burst varies considerably.

Remark 4.4 Some phenomenological, polynomial models for square-wave bursting have been proposed. See, for example, [11], [25], [7]. Analysis of models with two slow variables which exhibit square-wave bursting is given in [33].

Remark 4.5 A system of equations which give rise to square-wave bursting is [28]:

$$v' = -(g_{ca}m_\infty(v)(v - v_{ca}) + g_k w(v - v_k) + g_l(v - v_l) + g_{kca}z(y)(v - v_k)) + I$$

$$w' = 20\phi(w_\infty(v) - w)/\tau(v)$$

$$y' = 20\epsilon(-\mu g_{ca}m_\infty(v)(v - v_{ca}) - y)$$

where, $g_{ca} = 4.$, $g_k = 8.0$, $g_l = 2.0$, $v_k = -84$, $v_l = -60$, $v_{ca} = 120.0$, $I = 45$, $g_{kca} = .25$, $\phi = .23$, $\epsilon = .005$, and $\mu = .02$. The nonlinear functions are given by $m_\infty(v) = .5(1. + \tanh((v + 1.2)/18))$, $w_\infty(v) = .5(1. + \tanh((v - 12)/17.4))$, $z(y) = y/(1 + y)$ and $\tau(v) = \cosh((v - 12.)/34.8)$.

4.5 Traveling Wave Solutions

We have so far considered a model for neurons that ignores the spatial dependence. This has allowed us to study how temporal oscillatory behavior arises. One of the most important features of neurons is the propagating nerve impulse and this clearly requires consideration of spatial dynamics. The nerve impulse corresponds to a traveling wave solution and there has been extensive research on the mathematical mechanisms responsible for both the existence and stability of these types of solutions. Here we briefly illustrate how one constructs a traveling wave solution in a simple model, the FitzHugh-Nagumo equations. References related to this important topic are given later.

The FitzHugh-Nagumo equations can be written as:

$$v_t = v_{xx} + f(v) - w$$
$$w_t = \epsilon(v - \gamma w). \tag{23}$$

Here, (v, w) are functions of (x, t), $x \in R$ and $t \geq 0$. Moreover, $f(v) = v(1 - v)(v - a)$, $0 < a < 1/2$, ϵ is a small singular perturbation parameter, and γ is a positive constant chosen so that the curves $w = f(v)$ and $v = \gamma w$ intersect only at the origin.

A traveling wave solution of (23) is a solution of the form $(v(x,t), w(x,t)) = (V(z), W(z))$, $z = x + ct$; that is, a traveling wave solution corresponds to a solution that propagates with constant shape and velocity. The velocity is c as in not known a priori. We also assume that the traveling wave solution satisfies the boundary conditions $\lim_{z \to \pm\infty}(V(z), W(z)) = (0, 0)$.

Note that a traveling wave solution corresponds to a solution of the first order system of ordinary differential equations

$$V' = Y$$
$$Y' = cY - f(V) + W$$
$$W' = \frac{\epsilon}{c}(V - \gamma W) \qquad (24)$$

together with the boundary conditions

$$\lim_{z \to \pm\infty}(V(z), Y(z), W(z)) = (0, 0, 0) \qquad (25)$$

Hence, a traveling wave solution corresponds to a homoclinic orbit of a first order system. This homoclinic orbit will exist only for special values of the velocity parameter c.

One can use geometric singular perturbation methods, as described in Section 3.6, to construct a singular homoclinic orbit in which ϵ is formally set equal to zero. One needs to then rigorously prove that this singular solution perturbs to an actual homoclinic orbit that lies near the singular orbit for ϵ sufficiently small.

The singular orbit is constructed as follows. As before, the singular orbit consists of four pieces, as shown in Fig. 13. Two of these pieces correspond to the silent and active phases and the other two pieces correspond to the jump-up and jump-down between these phases. As before, we consider both fast and slow time scales.

The jump-up and jump-down pieces correspond to solutions of the fast equations. These are obtained by simply setting $\epsilon = 0$ in (24). The resulting equations are:

$$V' = Y$$
$$Y' = cY - f(V) + W$$
$$W' = 0 \qquad (26)$$

Note that W must be constant along this solution. For the jump-up (or front), we set $W \equiv 0$ and look for a solution of the first two equations of (24) that satisfy

$$\lim_{z \to -\infty}(V, Y) = (0, 0) \quad \text{and} \quad \lim_{z \to +\infty}(V, Y) = (1, 0) \qquad (27)$$

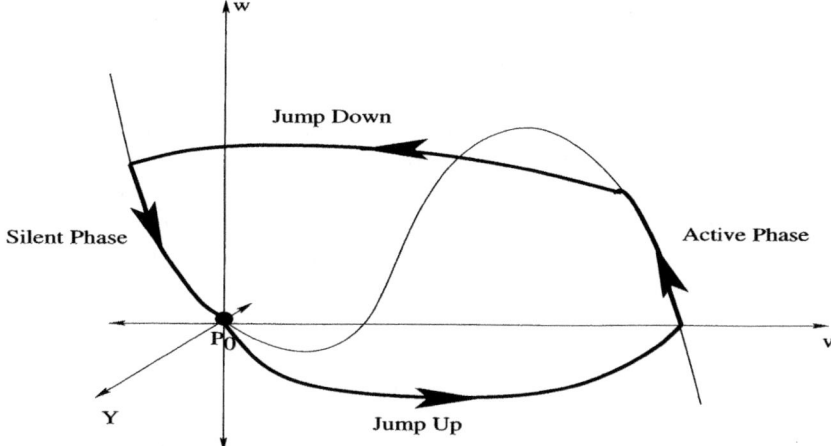

Fig. 13. The singular homoclinic orbit corresponding to a traveling wave solution.

It is well known that there exists a unique solution for a unique value of the parameter c. We denote this parameter as c_0. This is the velocity of the wave in the limit $\epsilon \to 0$.

For the jump-down (or back) we set $W \equiv W_0$, where W_0 is chosen so that if $c = c_0$ then there exists a solution of the first two equations in (26) such that $\lim_{z \to -\infty}(V, Y, W_0)$ lies along the right branch of the cubic $W = f(V)$ and $\lim_{z \to +\infty}(V, Y, W_0)$ lies along the left branch of this cubic. This is shown in Fig. 13. We note that W_0 is indeed uniquely determined.

We now consider the pieces of the singular traveling wave solution corresponding to the silent and active phases. We introduce the slow time scale $\eta = \epsilon z$ and then set $\epsilon = 0$ to obtain the slow equations

$$\begin{aligned} Y &= 0 \\ W &= f(V) \\ \dot{W} &= \frac{1}{c_0}(V - \gamma W) \end{aligned} \tag{28}$$

Here $\dot{W}$ corresponds to differentiation with respect to η. These equations demonstrate that during the silent and active phases, the singular solution lies along the cubic curve defined by $W = f(V)$, $Y = 0$. The complete singular homoclinic orbit is shown in Fig. 13.

Remark 4.6 References to rigorous studies of the existence and stability of traveling wave solutions can be found in [17].

5 Two Mutually Coupled Cells

5.1 Introduction

In this section, we consider a network consisting simply of two mutually coupled cells. By considering such a simple system, we are able to describe how we model networks of oscillators, the types of behavior that can arise in such systems and the mathematical techniques we use for the analysis of the behavior. For this discussion, we assume that each cell, without any coupling, is modeled as the relaxation oscillator

$$v' = f(v, w)$$
$$w' = \epsilon g(v, w) \tag{29}$$

Here ϵ is assumed to be small; that is, w represents a slowly evolving quantity. As in Section 3, we assume that the v-nullcline, $f(v, w) = 0$, defines a cubic-shaped curve and the w-nullcline, $g = 0$, is a monotone increasing curve which intersects $f = 0$ at a unique point p_0. We also assume that $f > 0$ ($f < 0$) below (above) the v-nullcline and $g > 0$ (< 0) below (above) the w-nullcline.

System (29) can be viewed as a simple model for a bursting neuron in which the active phase corresponds to the envelope of a burst's rapid spikes. Of course, a two-dimensional model for a single cell cannot exhibit the more exotic dynamics described in the previous section for a bursting cell. However, by considering a simple relaxation-type oscillator for each cell, we will be able to discuss how network properties contribute to the emergent behavior of a population of cells. It is, of course, a very interesting issue to understand how this population behavior changes when one considers more detailed models for each cell. Some results for more detailed models are given in [30].

In the next section, we describe how we model the two mutually coupled cells. The form of coupling used is referred to as *synaptic coupling* and is meant to correspond to a simple model for chemical synapses. There are many different forms of synaptic coupling. For example, it may be excitatory or inhibitory and it may exhibit either fast or slow dynamics. We are particularly interested in how the nature of the synaptic coupling affects the emergent population rhythm. A natural question is whether excitatory or inhibitory coupling leads to either synchronous or desynchronous rhythms. There are four possible combinations and all four may, in fact, be stably realized, depending on the details of the intrinsic and synaptic properties of the cells. Here we discuss conditions for when excitatory coupling leads to synchronous rhythms and inhibitory coupling leads to antiphase behavior.

5.2 Synaptic Coupling

We model a pair of mutually coupled neurons by the following system of differential equations

$$v_1' = f(v_1, w_1) - s_2 g_{syn}(v_1 - v_{syn})$$

$$w_1' = \epsilon g(v_1, w_1)$$

$$v_2' = f(v_2, w_2) - s_1 g_{syn}(v_2 - v_{syn}) \tag{30}$$

$$w_2' = \epsilon g(v_2, w_2)$$

Here (v_1, w_1) and (v_2, w_2) correspond to the two cells. The coupling term $s_j g_{syn}(v_i - v_{syn})$ can be viewed as an additional current which may change a cell's membrane potential v_i. The parameter g_{syn} corresponds to the maximal conductance of the synapse and is positive, while the reversal potential v_{syn} determines whether the synapse is excitatory or inhibitory. If $v < v_{syn}$ along each bounded singular solution, then the synapse is excitatory, while if $v > v_{syn}$ along each bounded singular solution, then the synapse is inhibitory.

The terms s_i, $i = 1, 2$, in (30) encode how the postsynaptic conductance depends on the presynaptic potentials v_i. There are several possible choices for the s_i. The simplest choice is to assume that $s_i = H(v_i - \theta_{syn})$, where H is the Heaviside step function and θ_{syn} is a threshold above which one cell can influence the other. Note, for example, that if $v_1 < \theta_{syn}$, then $s_1 = H(v_1 - \theta_{syn}) = 0$, so cell 1 has no influence on cell 2. If, on the other hand, $v_1 > \theta_{syn}$, then $s_1 = 1$ and cell 2 is affected by cell 1.

Another choice for the s_i is to assume that they satisfy a first order equation of the form

$$s_i' = \alpha(1 - s_i)H(v_i - \theta_{syn}) - \beta s_i \tag{31}$$

where α and β are positive constants and H and θ_{syn} are as before. Note that α and β are related to the rates at which the synapses turn on or turn off. For *fast synapses*, we assume that both of these constants are $O(1)$ with respect to ϵ. For a *slow synapse*, we assume that $\alpha = O(1)$ and $\beta = O(\epsilon)$; hence, a slow synapse activates on the fast time scale but turns off on the slow time scale.

5.3 Geometric Approach

All of the networks in this paper are analyzed by treating ϵ as a small, singular perturbation parameter. As in the previous section, the first step in the analysis is to identify the fast and slow variables. We then dissect the full system of equations into fast and slow subsystems. The fast subsystem is obtained by simply setting $\epsilon = 0$ in the original equations. This leads to a reduced set of equations for the fast variables with each of the slow variables held constant. The slow subsystems are obtained by first introducing the slow time scale $\tau = \epsilon t$ and then setting $\epsilon = 0$ in the resulting equations. This leads to a reduced system of equations for just the slow variables, after solving for each fast variable in terms of the slow ones. The slow subsystems determine the evolution of the slow variables while the cells are in either the active or the silent phase. During this time, each cell lies on either the left or the right branch of some "cubic" nullcline determined by the total synaptic input which the cell receives. This continues until one of the cells reaches the left or right "knee" of its corresponding cubic. Upon reaching a knee, the cell may either jump up from the silent to the active phase or jump down from the active to the silent phase. The jumping up or down process is governed by the fast equations.

For a concrete example, consider two mutually coupled cells with fast synapses. The dependent variables (v_i, w_i, s_i), $i = 1, 2$, then satisfy (30) and (31). The slow equations are

$$0 = f(v_i, w_i) - s_j g_{syn}(v_i - v_{syn})$$

$$\dot{w}_i = g(v_i, w_i) \tag{32}$$

$$0 = \alpha(1 - s_i)H(v_i - \theta_{syn}) - \beta s_i$$

where differentiation is with respect to τ and $i \neq j$. The first equation in (32) states that (v_i, w_i) lies on a curve determined by s_j. The third equation states that if cell i is silent ($v_i < \theta_{syn}$), then $s_i = 0$, while if cell i is active, then $s_i = \frac{\alpha}{\alpha+\beta} \equiv s_A$. We demonstrate that it is possible to reduce (32) to a single equation for each of the slow variables w_i. Before doing this, it will be convenient to introduce some notation.

Let $\Phi(v, w, s) \equiv f(v, w) - g_{syn}s(v - v_{syn})$. If g_{syn} is not too large, then each $C_s \equiv \{\Phi(v, w, s) = 0\}$ defines a cubic-shaped curve. We express the left and right branches of C_s by $\{v = \Phi_L(w, s)\}$ and $\{v = \Phi_R(w, s)\}$, respectively. Finally, let

$$G_L(w, s) = g(\Phi_L(w, s), w) \quad \text{and} \quad G_R(w, s) = g(\Phi_R(w, s), w)$$

Now the first equation in (32) can be written as $0 = \Phi(v_i, w_i, s_j)$ with s_j fixed. Hence, $v_i = \Phi_\alpha(w_i, s_j)$ where $\alpha = L$ if cell i is silent and $\alpha = R$ if cell i is active. It then follows that each slow variable w_i satisfies the single equation

$$\dot{w}_i = G_\alpha(w_i, s_j) \tag{33}$$

By dissecting the full system into fast and slow subsystems, we are able to construct singular solutions of (30),(31). In particular, this leads to sufficient conditions for when there exists a singular synchronous solution and when this solution is (formally) asymptotically stable. The second step in the analysis is to rigorously prove that the formal analysis, in which $\epsilon = 0$, is justified for small $\epsilon > 0$. This raises some very subtle issues in the geometric theory of singular perturbations, some of which have not been completely addressed in the literature. For most of the results presented here, we only consider singular solutions.

We note that the geometric approach used here is somewhat different from that used in many dynamical systems studies (see, for example, [28]). All of the networks considered here consist of many differential equations, especially for larger networks. Traditionally, one would interpret the solution of this system as a single trajectory evolving in a very large dimensional phase space. We consider several trajectories, one corresponding to a single cell, moving around in a much lower dimensional phase space (see also [43], [41], [34], [40], [30]). After reducing the full system to a system for just the slow variables, the dimension of the lower dimensional phase space equals the number of slow intrinsic variables and slow synaptic variables corresponding to each cell. In the worst case considered here, there is only one slow intrinsic variable for each cell and one slow synaptic variable; hence, we never have to consider phase spaces with dimension more than two. Of course, the particular phase space we need to consider may change, depending on whether the cells are active or silent and also depending on the synaptic input that a cell receives.

5.4 Synchrony with Excitatory Synapses

Consider two mutually coupled cells with excitatory synapses. Our goal here is to give sufficient conditions for the existence of a synchronous solution and its stability. Note that if the synapses are excitatory, then the curve $C_A \equiv C_{s_A}$ lies 'above' $C_0 \equiv \{f = 0\}$ as shown in Fig. 14. This is because for an excitatory synapse, $v < v_{syn}$ along the synchronous solution. Hence, on C_A, $f(v, w) = g_{syn} s_A (v - v_{syn}) < 0$, and we are assuming that $f < 0$ above C_0. If g_{syn} is not too large, then both C_0 and C_A will be cubic shaped. We assume that the threshold θ_{syn} lies between the two knees of C_0. In the statement of the following result, we denote the left knee of C_0 by (v_{LK}, w_{LK}).

Theorem: Assume that each cell, without any coupling, is oscillatory. Moreover, assume the synapses are fast and excitatory. Then there exists a synchronous periodic solution of (30), (31). This solution is asymptotically stable if one of the following two conditions is satisfied.

(H1) $\frac{\partial f}{\partial w} < 0$, $\frac{\partial g}{\partial v} > 0$, and $\frac{\partial g}{\partial w} < 0$ near the singular synchronous solution.

(H2) $|g(v_{LK}, w_{LK})|$ is sufficiently small.

Remark 5.1 We note that the synchronous solution cannot exist if the cells are excitable and the other hypotheses, concerning the synapses, are satisfied. This is because along a synchronous solution, each (v_i, w_i) lies on the left branch of C_0 during the silent phase. If the cells are excitable, then each (v_i, w_i) will approach the point where the w-nullcline $\{g = 0\}$ intersects the left branch of C_0. The cells, therefore, will not be able to jump up to the active phase.

Remark 5.2 The assumptions concerning the partial derivatives of f and g in (H1) are not very restrictive since we are already assuming that $f > 0$ (< 0) below (above) the v-nullcline and $g > 0$ (< 0) below (above) the w-nullcline.

Remark 5.3 A useful way to interpret (H2) is that the silent phases of the cells are much longer than their active phases. This is because $g(v_{LK}, w_{LK})$ gives the rate at which the slow variables w_i evolve near the end of the silent phase. Note that $g(v_{LK}, w_{LK})$ will be small if the left knee of C_0 is very close to the w-nullcline.

Proof: We first consider the existence of the synchronous solution. This is straightforward because along a synchronous solution $(v_1, w_1, s_1) = (v_2, w_2, s_2) \equiv (v, w, s)$ satisfy the reduced system

$$v' = f(v, w) - sg_{syn}(v - v_{syn})$$

$$w' = \epsilon g(v, w)$$

$$s' = \alpha(1 - s)H(v - \theta_{syn}) - \beta s$$

The singular solution consists of four pieces. During the silent phase, $s = 0$ and (v, w) lies on the left branch of C_0. During the active phase $s = s_A$ and (v, w) lies on the right branch of C_A. The jumps between these two phases occur at the left and right knees of the corresponding cubics.

We next consider the stability of the synchronous solution to small perturbations. We begin with both cells close to each other in the silent phase on the left branch of C_0, with cell 1 at the left knee ready to jump up. We follow the cells around in phase space by constructing the singular solution until one of the cells returns to the left knee of C_0. As before, the singular solution consists of four pieces. We need to show that the cells are closer to each other after this complete cycle than they were initially.

The first piece of the singular solution begins when cell 1 jumps up. When $v_1(t)$ crosses θ_{syn}, $s_1(t) \to s_A$. This raises the cubic corresponding to cell 2 from C_0 to C_A. If $|w_1(0) - w_2(0)|$ is sufficiently small, corresponding to a sufficiently small perturbation, then cell 2 lies below the left knee of C_A. The fast equations then force cell 2 to also jump up to the active phase, as shown in Fig. 14. Note that this piece takes place on the fast time scale. Hence, on the slow time scale, both cells jump up at precisely the same time.

During the second piece of the singular solution, both oscillators lie in the active phase on the right branch of C_A. Note that the ordering in which the oscillators track along the left and right branches has been reversed. While in the silent phase, cell 1 was ahead of cell 2. In the active phase, cell 2 leads the way. The oscillators remain on the right branch of C_A until cell 2 reaches the right knee.

The oscillators then jump down to the silent phase. Cell 2 is the first to jump down. When $v_2(t)$ crosses θ_{syn}, s_2 switches from s_A to 0 on the fast time scale. This lowers the cubic corresponding to cell 1 from C_A to C_0. If, at this time, cell 1 lies above the right knee of C_A, then cell 1 must jump down to the silent phase. This will certainly be the case if the cells are initially close enough to each other.

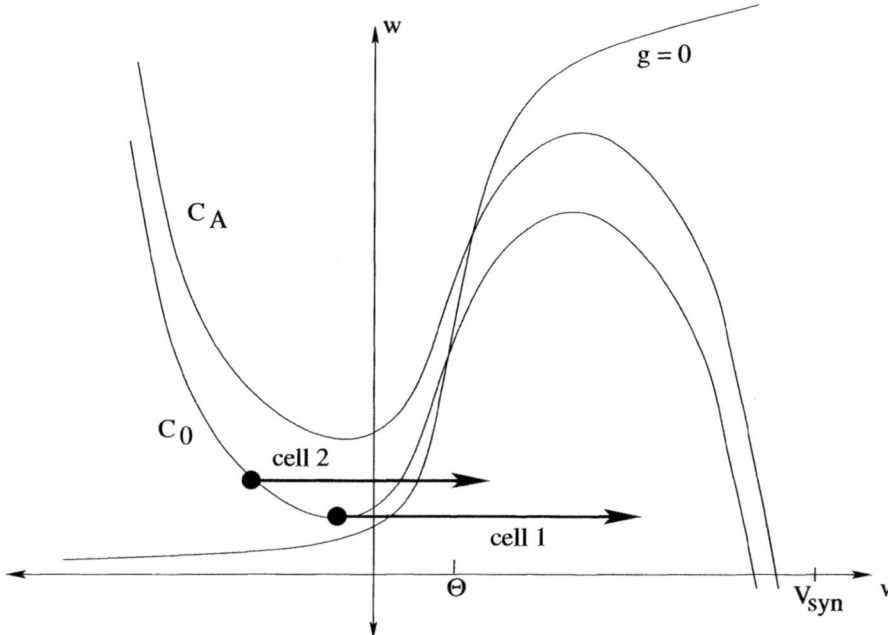

Fig. 14. Nullclines for an oscillatory relaxation oscillator with (C_A) and without (C_0) excitatory coupling. Note that cell 2 responds to cell 1 through Fast Threshold Modulation.

During the final piece of the singular solution, both oscillators move down the left branch of C_0 until cell 1 reaches the left knee. This completes one full cycle.

To prove that the synchronous solution is stable, we must show that the cells are closer to each other after this cycle; that is, there is compression in the distance between the cells. There are actually several ways to demonstrate this compression; these correspond to two different ways to define what is meant by the 'distance' between the cells. Here we consider a Euclidean metric, which is defined as follows: Suppose that both cells lie on the same branch of the same cubic and the coordinates of cell i are (v_i, w_i). Then the distance between the cells is defined as simply $|w_1 - w_2|$. Note that during the jump up and the jump down, this metric remains invariant. This is because the jumps are horizontal so the values of w_i do not change. If there is compression, therefore, it must take place as the cells evolve in the silent and active phases. We now show that this is indeed the case if (H1) is satisfied.

Suppose that when $\tau = 0$, both cells lie in the silent phase on C_0. We assume, for convenience, that $w_2(0) > w_1(0)$. We need to prove that $w_2(\tau) - w_1(\tau)$ decreases as long as the cells remain in the silent phase. Now each w_i satisfies (33) with $\alpha = L$ and $s_j = 0$. Hence,

$$w_i(\tau) = w_i(0) + \int_0^\tau G_L(w_i(\xi), 0)d\xi$$

and, using the Mean Value Theorem,

$$
\begin{aligned}
w_2(\tau) - w_1(\tau) &= w_2(0) - w_1(0) \\
&\quad + \int_0^\tau G_L(w_2(\xi), 0) - G_L(w_1(\xi), 0) \, d\xi \\
&= w_2(0) - w_1(0) \\
&\quad + \int_0^\tau \frac{\partial G_L}{\partial w}(w^*, 0)(w_2(\xi) - w_1(\xi))d\xi
\end{aligned}
\tag{34}
$$

for some w^*. Now $G_L(w, s) = g(\Phi_L(w), w)$. Hence, $\frac{\partial G_L}{\partial w} = g_v \Phi'_L(w) + g_w$. We assume in (H1) that $g_v > 0$ and $g_w < 0$ near the synchronous solution. Moreover, $\Phi'_L(w) < 0$ because $v = \Phi_L(w)$ defines the left branch of the cubic C_0 which has negative slope. It follows that $\frac{\partial G_L}{\partial w} < 0$, and therefore, from (34), $w_2(\tau) - w_1(\tau) < w_2(0) - w_1(0)$. This gives the desired compression; a similar computation applies in the active phase. We note that if there exists $\gamma > 0$ such that $\frac{\partial G_L}{\partial w} < -\gamma$ along the left branch, then Gronwall's inequality shows that $w_2(\tau) - w_1(\tau)$ decreases at an exponential rate.

We next consider (H2) and demonstrate why this leads to compression of trajectories. Suppose, for the moment, that $g(v_{LK}, w_{LK}) = 0$; that is, the left knee of C_0 touches the w-nullcline at some fixed point. Then both cells will approach this fixed point as they evolve along the left branch of C_0 in the silent phase. There will then be an infinite amount of compression, since both cells approach the same fixed point. It follows that we can assume that the compression is as large as we please by making $g(v_{LK}, w_{LK})$ sufficiently small. If the compression is sufficiently large, then it will easily dominate any possible expansion over the remainder of the cells' trajectories. This will, in turn, lead to stability of the synchronous solution.

Remark 5.4 The mechanism by which one cell fires, and thereby raises the cubic of the other cell such that it also fires, was referred to as *Fast Threshold Modulation (FTM)* in [34]. There, a time metric was introduced to establish the compression of trajectories of excitatorily coupled cells, which implies the stability of the synchronous solution. A detailed discussion of the time metric can be found in [20]; see also [23].

Remark 5.5 While the synchronous solution has been shown to be stable, it need not be globally stable. In [21], it is shown that this network may exhibit stable antiphase solutions if certain assumptions on the parameters and nonlinear functions are satisfied.

We have so far considered a completely homogeneous network with just two cells. The analysis generalizes to larger inhomogeneous networks in a straightforward manner, if the degree of heterogeneity between the cells is not too large. The major difference in the analysis is that, with heterogeneity, the cells may lie on different branches of different cubics during the silent and active phases. The resulting solution cannot be perfectly synchronous; however, as demonstrated in [43], one can often expect synchrony in the jump-up, but not in the jump-down. Related work on heterogeneous networks include [35], [26], [4].

One may also consider, for example, an arbitrarily large network of identical oscillators with nearest neighbor coupling. We do not assume that the strength of coupling is homogeneous. Suppose that we begin the network with each cell in the silent phase. If the cells are identical, then they must all lie on the left branch of C_0. Now if one cell jumps up it will excite its neighbors and raise their corresponding cubics. If the cells begin sufficiently close to each other, then these neighbors will jump up due to FTM. In a similar manner, the neighbor's neighbors will also jump due to FTM and so on until every cell jumps up. In this way, every cell jumps up at the same (slow) time. While in the active phase, the cells may receive different input and, therefore, lie on the right branches of different cubics. Once one of the cells jumps down, there is no guarantee that other cells will also jump down at this (slow) time, because the cells to which it is coupled may still receive input from other active cells. Hence, one cannot expect synchrony in the jumping down process. Eventually every cell must jump down. Note that there may be considerable expansion in the distance between the cells in the jumping down process. If $|g(v_{LK}, w_{LK})|$ is sufficiently small, however, as in the previous result, then there will be enough compression in the silent phase so that the cells will still jump up together. Here we assumed that the cells are identical; however, the analysis easily follows if the heterogeneities among the cells are not too large. A detailed analysis of this network is given in [43].

5.5 Desynchrony with Inhibitory Synapses

We now consider two mutually coupled cells with inhibitory synapses. Under this coupling, the curve C_A now lies below C_0. As before, we assume that g_{syn} is not too large, such that both C_0 and C_A are cubic shaped. We also assume that the right knee of C_A lies above the left knee of C_0 as shown in Fig. 15. Some assumptions on the threshold θ_{syn} are also required. For now, we assume that θ_{syn} lies between the left knee of C_0 and right knee of C_A.

We will assume throughout this section that the synapses are fast and inhibitory. The main results state that if a synchronous solution exists then it must be unstable. The network will typically exhibit either out-of-phase oscillations or a completely quiescent state and we give sufficient conditions for when either of these arises. We note that the network may exhibit bistability; both the out-of-phase and completely quiescent solutions may exist and be stable for the same parameter values. These results are all for singular solutions. Some rigorous results for $\epsilon > 0$ are given in [41].

The first result concerns the existence and stability of the synchronous solution.

Theorem: Assume that the synapses are fast and inhibitory. If each cell, without any coupling, is oscillatory and θ_{syn} is sufficiently large, then there exists a singular synchronous solution. This solution is unstable. If each cell, without any coupling, is excitable, then there does not exist a singular synchronous solution.

Proof: The existence of a singular synchronous solution for oscillatory cells follows precisely as in the previous section. During the silent phase, the trajectory lies on the left branch of C_0, while in the active phase it lies on the right branch of C_A. Note that we require that the right knee of C_A lies above the left knee of C_0. Moreover, when the synchronous solution jumps up and crosses the threshold $v = \theta_{syn}$, it should lie to the right of the middle branch of C_A; otherwise, it would fall down to the silent phase. This is why we assume that θ_{syn} is sufficiently large.

This solution is unstable for the following reason. Suppose both cells are initially very close to each other on C_0. The cells then evolve on C_0 until one of the cells, say cell 1, reaches the left knee of C_0. Cell 1 then jumps up to the active phase. When v_1 crosses the threshold θ_{syn}, s_1 switches from 0 to s_A and cell 2 jumps from C_0 to C_A, as shown in Fig. 15. This demonstrates that the cells are uniformly separated for arbitrarily close initial data. The synchronous solution must, therefore, be unstable.

The synchronous solution cannot exist if the cells are excitable for precisely the same reason discussed in the previous section. If such a solution did exist then each cell would lie on C_0 during its silent phase. Each cell would then approach the stable fixed point on this branch and would never be able to jump up to the active phase.

We next consider out-of-phase oscillatory behavior. One interesting feature of mutually coupled networks is that such oscillations can arise even if each cell is excitable for fixed levels of synaptic input. The following theorem gives sufficient conditions for when this occurs. We will require that the active phase of the oscillation is sufficiently long. To give precise conditions, we introduce the following notation.

Assume that the left and right knees of $C_0 \cdot$ are at (v_{LK}, w_{LK}) and (v_{RK}, w_{RK}), respectively. If the w-nullcline intersects the left branch of C_A, then we denote this point by $(v_A, w_A) = p_A$. We assume that $w_A < w_{LK}$. Let τ_L be the (slow) time it takes for the solution of (33) with $\alpha = L$ and $s = s_A$ to go from $w = w_{RK}$ to $w = w_{LK}$, and let τ_R be the time it takes for the solution of (33) with $\alpha = R$ and $s = 0$ to go from $w = w_{LK}$ to $w = w_{RK}$. Note that τ_L is related to the time a solution spends in the silent phase, while τ_R is related to the time a solution spends in the active phase.

Theorem: Assume that the cells are excitable for each fixed level of synaptic input and the synapses are fast, direct, and inhibitory. Moreover, assume

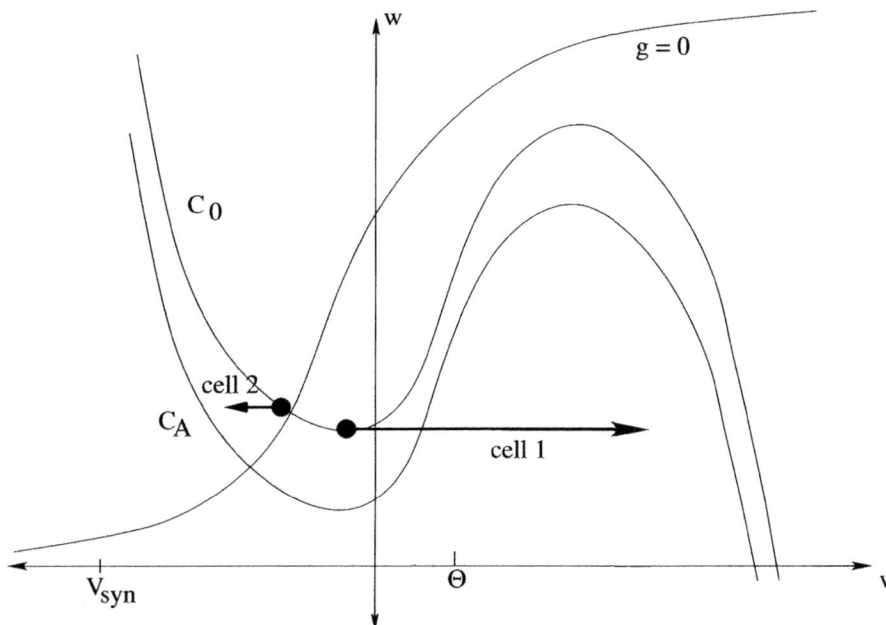

Fig. 15. Instability induced by mutual inhibition. Cell 2 jumps to C_A when cell 1 fires.

that $w_A < w_{LK}$ and $\tau_L < \tau_R$. Then the network exhibits stable out-of-phase oscillatory behavior.

Remark 5.6 We do not claim that the out-of-phase solution is uniquely determined or that it corresponds to antiphase behavior. These results may hold; however, their proofs require more analysis than that given here.

Remark 5.7 The rest state with each cell at the fixed point on C_0 also exists and is stable. Hence, if the hypotheses of the last Theorem are satisfied, then the network exhibits bistability.

Proof: Suppose that we begin with cell 1 at the right knee of C_0 and cell 2 on the left branch of C_A with $w_A < w_2(0) < w_{LK}$. Then cell 1 jumps down and, when v_1 crosses the threshold θ_{syn}, cell 2's cubic switches from C_A to C_0. Since $w_2(0) < w_{LK}$, cell 2 lies below the left knee of C_0, so it must jump up to the active phase. After these jumps, cell 1 lies on the left branch of C_A, while cell 2 lies on the right branch of C_0.

Cell 2 then moves up the right branch of C_0 while cell 1 moves down the left branch of C_A, approaching p_A. This continues until cell 2 reaches the right knee of C_0 and jumps down. We claim that at this time, cell 1 lies below the left knee of C_0, so it must jump up. We can then keep repeating this argument to obtain the sustained out-of-phase oscillations. The reason why cell 1 lies below the left knee of C_0 when cell 2 jumps down is because it spends a sufficiently long amount of time in the silent phase. To estimate this time, note that because cell 2 was initially below the left knee of C_0, the time it spends in the active phase before jumping down is greater than τ_R. Hence, the time cell 1 spends in the silent phase from the time it jumps down is greater than $\tau_R > \tau_L$. From the definitions, since cell 1 was initially at the right knee of C_0, it follows that cell 1 must be below the left knee of C_0 when cell 2 jumps down, which is what we wished to show.

Remark 5.8 Wang and Rinzel [45] distinguish between "escape" and "release" in producing out-of-phase oscillations. In the proof of the preceding theorem, the silent cell can only jump up to the active phase once the active cell jumps down and releases the silent cell from inhibition. This is referred to as the release mechanism and is often referred to as *post inhibitory rebound* [9]. To describe the escape mechanism, suppose that each cell is oscillatory for fixed levels of synaptic input. Moreover, one cell is active and the other is inactive. The inactive cell will then be able to escape the silent phase from the left knee of its cubic, despite the inhibition it receives from the active cell. Note that when the silent cell jumps up, it inhibits the active cell. This lowers the cubic of the active cell, so it may be forced to jump down before reaching a right knee.

Remark 5.9 We have presented rigorous results that demonstrate that excitation can lead to synchrony and inhibition can lead to desynchrony. These results depended on certain assumptions, however. We assumed, for example, that the synapses turned on and off on a fast time scale; moreover, the results hold in some singular limit. In is, in fact, possible for excitatory synapses to generate stable desynchronous oscillations and for inhibitory synapses to generate stable synchronous oscillations. Conditions for when these are possible has been the subject of numerous research articles. References may be found in [40].

Remark 5.10 The two cell model (30) can generate other rhythms besides those discussed so far. For example, it is possible that one cell periodically fires action potentials, while the other cell is always silent. This is sometimes refereed to as a *suppressed solution*. It arises if the rate at which inhibition turns off is slower than the rate at which a cell recovers in the silent phase. That is, suppose cell 1 fires. This sends inhibition to cell 2, preventing it from firing. Now if cell 1 is able to recover from its silent phase before the inhibition to cell 2 wears off, then cell 1 will fire before cell 2 is able to. This will complete one cycle and cell 2 will continue to be suppressed. If the time for cells to recover in the silent phase is comparable to the time for inhibition to decay, then more exotic solutions are possible (see [40]).

6 Activity Patterns in the Basal Ganglia

6.1 Introduction

In this final chapter, we discuss a recent model for neuronal activity patterns in the basal ganglia [42]. This is a part of the brain believed to be involved in the control of movement. Dysfunction of the basal ganglia is associated with movement disorders such as Parkinson's disease and Huntington's disease. Neurons within the basal ganglia are also the target of recent surgical procedures, including deep brain stimulation. An introduction to the basal ganglia can be found in [18].

The issues discussed arise in numerous other neuronal systems. We shall describe how these large neuronal networks are modeled, what population rhythms may arise in these networks and the possible roles of these activity patterns.

6.2 The Basal Ganglia

The basal ganglia consist of several nuclei; these are illustrated in Fig. 16. The primary input nucleus is the striatum; it receives motor information from the cortex. The primary output nuclei are the internal segment of the globus pallidus (GPi) and the substantia nigra par retularis (SNr). Neuronal information passes through the basal ganglia through two routes. The direct pathway passes directly from the striatum to the output nuclei. In the indirect pathway, the information passes from the striatum to the external segment of the globus pallidus (GPe) onto the subthalamic nuclues (STN) and then onto the output nuclei. The final nucleus is the substantia nigra par compacta (SNc). This is the primary source of dopamine.

Fig. 16 illustrates that some of the pathways within the basal ganglia are excitatory and some are inhibitory. Most of the pathways are inhibitory except for those that originate in the STN. Note that the pathways arising from SNc are labeled both inhibitory and excitatory. This is because there are different classes of dopamine receptors within the striatum. We also note that Fig. 16 illustrates only some of the pathways reported in the literature.

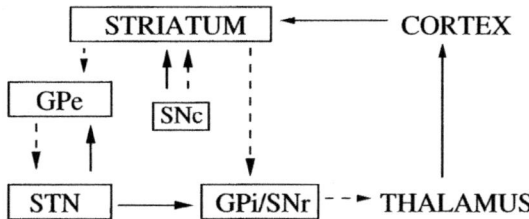

Fig. 16. Nuclei within the basal ganglia. Solid arrows indicate excitatory connections and dashed arrows indicate inhibitory connections.

Parkinson's disease is associated with a severe reduction of dopamine. Experiments have also demonstrated that during Parkinson's disease, there is a change in the neuronal activity of the output nucleus GPi. Previous explanations for how a loss of dopamine leads to altered neuronal activity in GPi have been in terms of an average firing rate of neurons; that is, the average number of action potentials in some fixed interval of time. A detailed description of this explanation can be found in [1, 8, 42]. It has been successful in accounting for some features of PD; however, it cannot account for others. For example, it is not at all clear how one can explain tremor. It also cannot account for recent experiments that demonstrate that there is an increased level synchronization among neurons in the STN and GPi during a parkinsonian state [14]. Several authors have suggested that the pattern of neuronal activity, not just the firing rate, is crucially important.

The goal of the modeling study in [42] is to test hypotheses on how the loss of dopamine may lead to tremor-like oscillations and changes in firing patterns. We construct a model for neurons within GPe and STN based on recent experiments [2]. We use computational methods to study the types of activity patterns that arise in this model. In particular, we demonstrate that the model can exhibit irregular uncorrelated patterns, synchronous tremor-like rhythms, and propagating wave-like activity. In the next subsection, we describe the computational model. We then describe the types of activity patterns that arise in the model.

6.3 The Model

Here we describe the model for the STN and GPe network. The detailed equations are given in [42]. These equations are derived using the Hodgkin-Huxley formalism discussed earlier. The precise equations are different from the Hodgkin-Huxley equations, however. This is because the STN and GPe neurons contain channels different from those in the squid's giant axon. In particular, calcium plays a very important role in generating the action potential of STN and GPe neurons. There are two types of potassium channels, one of which depends on the concentration of intracellular calcium (along with membrane potential). There are also two types of calcium channels in STN neurons.

The membrane potential of each STN neuron obeys the current balance equation:

$$C_m \frac{dV}{dt} = -I_L - I_K - I_{Na} - I_T - I_{Ca} - I_{AHP} - I_{G \to S} .$$

The leak current is given by $I_L = g_L(v - v_L)$, and the other voltage-dependent currents are described by the Hodgkin-Huxley formalism as follows: $I_K = g_K n^4(v - v_K)$, $I_{Na} = g_{Na} m_\infty^3(v) h(v - v_{Na})$, $I_T = g_T a_\infty^3(v) b_\infty^2(r)(v - v_{Ca})$, and $I_{Ca} = g_{Ca} s_\infty^2(v)(v - v_{Ca})$. The slowly-operating gating variables n,

h, and r are treated as functions of both time and voltage, and have first order kinetics governed by differential equations of the form $\frac{dX}{dt} = \phi_X \frac{(X_\infty(v) - X)}{\tau_X(v)}$ (where X can be n, h, or r), with $\tau_X(v) = \tau_X^0 + \frac{\tau_X^1}{1 + \exp[-(v - \theta_X^\tau)/\sigma_X^\tau]}$. Activation gating for the rapidly activating channels (m, a, and s) was treated as instantaneous. For all gating variables $X = n, m, h, a, r$, or s, the steady state voltage dependence was determined using $X_\infty(v) = \frac{1}{1 + \exp[-(v - \theta_X)/\sigma_X]}$. The gating variable b was modeled in a similar, but somewhat different, manner; we do not describe this here. As the final intrinsic current, we take $I_{AHP} = g_{AHP}(v - v_K)\frac{[Ca]}{([Ca] + k_1)}$ where $[Ca]$, the intracellular concentration of Ca^{2+} ions, is governed by $[Ca]' = \epsilon(-I_{Ca} - I_T - k_{Ca}[Ca])$.

The current $I_{G \to S}$ that represents synaptic input from the GPe to STN is modeled as $I_{G \to S} = g_{G \to S}(v - v_{G \to S})\sum s_j$. The summation is taken over the presynaptic GPe neurons, and each synaptic variable s_j solves a first order differential equation $s_j' = \alpha H_\infty(vg_j - \theta_g)(1 - s_j) - \beta s_j$. Here vg_j is the membrane potential of the GPe neuron j, and $H_\infty(v) = 1/(1 + \exp[-(v - \theta_g^H)/\sigma_g^H])$.

The precise forms of the nonlinear functions in this model, along with parameter values, are given in [42]. The GPe neurons are modeled in a similar way. We do not describe these equations here.

The model STN neurons were adjusted to exhibit properties that are characteristic of the firing of STN neurons in experiments [2]. Fig. 17, left column, shows the firing properties of the model STN neurons. These cells fire intrinsically at approximately 3 Hz and exhibit high frequency sustained firing and strong rebound bursts after release from hyperpolarizing current. Fig. 17, right column, illustrates the firing properties of single GPe neurons. These cells can fire rapid periodic spikes with sufficient applied current. They also display bursts of activity when subjected to a small constant hyperpolarizing current.

Currently the details of connections between STN and GPe cells are poorly understood. It is known that STN neurons provide one of the largest sources of excitatory input to the globus pallidus and that the GPe is a major source of inhibitory afferents to the STN. However, the spatial distribution of axons in each pathway, as well as the number of cells innervated by single neurons in each direction, are not known to the precision required for a computer model. Therefore, in [42] we consider multiple architectures in order to study what types of activity patterns may arise in a particular class of network architecture. In the model networks, each GPe neuron sends inhibition to other GPe neurons as well as to one or more STN neurons. Each STN neuron sends excitation to one or more GPe neurons. A prototype network is illustrated in Figure 18.

6.4 Activity Patterns

Two activity patterns displayed by the model are shown in Fig. 19. The left column displays irregular and weakly correlated firing of each cell. The volt-

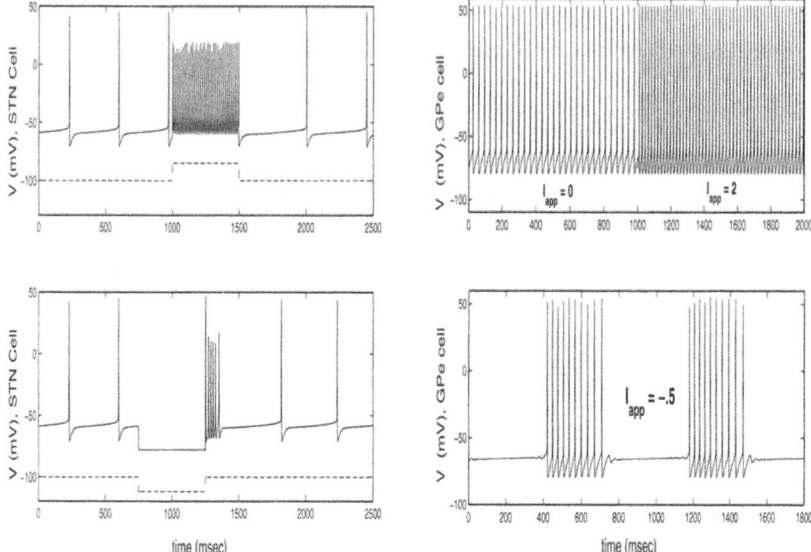

Fig. 17. Voltage traces for (left) STN and (right) GPe neurons for different levels of applied current. STN cells display high frequency sustained firing with higher input (as shown by the elevated dashed line) and fire rebound bursts after release from hyperpolarizing current. GPe cells fire rapid periodic spikes for positive input and fire bursts of spikes for small negative applied current.

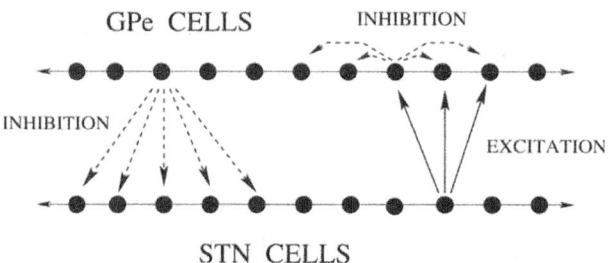

Fig. 18. Architecture of the STN/GPe network.

age traces of two STN neurons are shown. shown. Irregular activity arises in sparsely connected, unstructured networks in which each neuron is synaptically connected to only a small number of other neurons chosen at random. It is possible for irregular activity to also arise in structured networks, however.

The right column of Fig. 18 displays clustered activity, in which each structure is divided into subsets of neurons that become highly correlated with each other. The most commonly observed clustered pattern consists of two clusters, with alternating pairs of cells belonging to opposite clusters. Different clusters alternate firing, and in this pattern, cluster membership is persistent

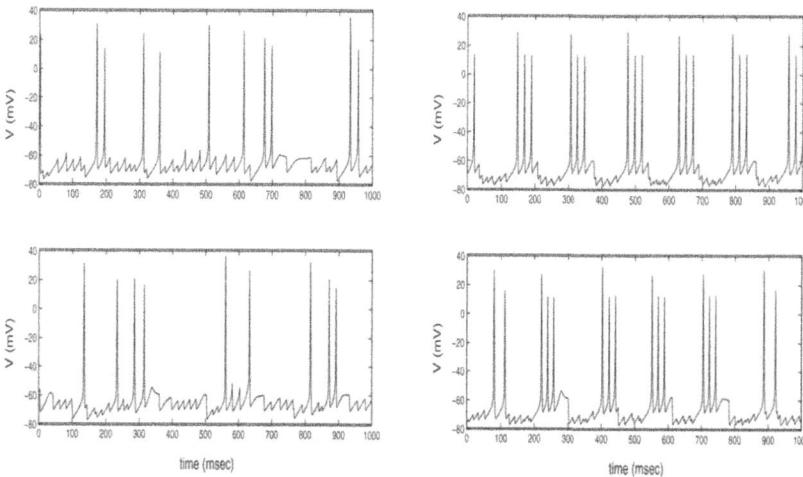

Fig. 19. Irregular and clustered patterns. Each column shows the voltage traces of two STN neurons.

over time. Both episodic and continuous clustering are possible. Clustered activity typically arises in networks with a structured, sparsely connected architecture.

A third type of activity pattern is traveling waves (not shown). Propagating wave-like activity can be very robust and exist over a wide range of parameter values. They typically arise in networks with a structured, tightly connected architecture. The speed of the wave depends on both the footprint of network architecture and both intrinsic and synaptic time-scales.

We note that both of the patterns shown in Fig. 19 are generated for a network with exactly the same architecture. In order to switch from the irregular pattern to the synchronous pattern, we increase the applied current to the GPe cells (this corresponds to input from the striatum) and the level of intra-GPe inhibition.

6.5 Concluding Remarks

We have shown that in a biophysical, conductance-based model that the cellular properties of the STN and GPe cells can give rise to a variety of rhythmic or irregular self-sustained firing patterns, depending on both the arrangement of connections among and within the nuclei and the effective strengths of the connections. The dependence on network architecture points out the importance of certain missing pieces of anatomical information. It is important to determine the precision of the spatial organization of connections between the STN and GPe neurons and whether the two nuclei project on each other in a reciprocal or out of register manner.

According to recent studies, correlated oscillatory activity in the GPe and STN neurons is closely related to the generation of the symptoms of Parkinsonism. Previous firing rate models hold that during Parkinsonian states, an increased level of inhibition from the striatum to GPe causes a decrease in the activity of GPe. This in turn would send less inhibition to STN, thus increasing STN activity and ultimately leading to increased inhibitory output from the basal ganglia to the thalamus. In our model network, a more complex picture emerges, in which the STN and GPe are spontaneously oscillatory and synchronous, whereas intra-GPe inhibition and appropriate level of input from the striatum can act to suppress rhythmic behavior.

The analysis described in earlier chapters is extremely useful in understanding the mechanisms responsible for the generation of the different firing patterns arising in the STN/GPe model. The simple two-cell models considered earlier illustrate, for example, that inhibition may play multiple roles in the generation of activity patterns. In the clustered rhythm, for example, active STN neurons need moderate levels of feedback inhibition from GPe to synchronize among themselves. Silent STN neurons, on the other hand, are prevented from firing because they receive more powerful tonic inhibition. For the generation of propagating waves, intra-GPe inhibition is needed to prevent activity from persisting in the wake of the wave. Hence, this inhibition helps to organize the network into a structured activity pattern. If one increases the intra-GPe inhibition, this can desynchronize the GPe oscillations and irregular firing may result.

The STN/GPe model is an example of an excitatory-inhibitory network. This type of model arises in other neuronal systems. For example, recent models for thalamic sleep rhythms share many of the properties of the STN/GPe model. References to papers on the thalamic sleep rhythms can be found in [31].

References

1. R.L. Albin, A.B. Young, and J.B. Penney. The functional anatomy of basal ganglia disorders. *Trends in Neurosci.*, **12**:366–375, 1989.
2. M.D. Bevan and C.J. Wilson. Mechanisms underlying spontaneous oscillation and rhythmic firing in rat subthalamic neurons. *J. Neurosci.*, **19**:7617–7628, 1999.
3. R.J. Butera, J. Rinzel, and J.C. Smith. Models of respiratory rhythm generation in the pre-Botzinger complex: I. Bursting pacemaker model. *J. Neurophysiology*, **82**:382–397, 1999.
4. R.J. Butera, J. Rinzel, and J.C. Smith. Models of respiratory rhythm generation in the pre-Botzinger complex: II. Populations of coupled pacemaker neurons. *J. Neurophysiology*, **82**:398–415, 1999.
5. T.R Chay and J. Keizer. Minimal model for membrane oscillations in the pancreatic β-cell. *Biophys. J.*, **42**:181–190, 1983.
6. T.R. Chay and J. Rinzel. Bursting, beating, and chaos in an excitable membrane model. *Biophys. J.*, **47**:357–366, 1985.

7. G. de Vries and R.M. Miura. Analysis of a class of models of bursting electrical activity in pancreatic β-cells. *SIAM J. Appl. Math.*, **58**:607–635, 1998.
8. M.R. DeLong. Activity of pallidal neurons during movement. *J. Neurophysiol.*, **34**:414–427, 1971.
9. W. Friesen. Reciprocal inhibition, a mechanism underlying oscillatory animal movements. *Neurosci. Behavior*, **18**:547–553, 1994.
10. D. Golomb, X.-J. Wang, and J. Rinzel. Synchronization properties of spindle oscillations in a thalamic reticular nucleus model. *J. Neurophysiol.*, **72**:1109–1126, 1994.
11. J.L Hindmarsh and R.M. Rose. A model of neuronal bursting using three coupled first order differential equations. *Proc. R. Soc. Lond., Ser. B*, **221**:87–102, 1984.
12. A.L. Hodgkin and A.F. Huxley. A quantitative description of membrane current and its application to conduction and excitation in a nerve. *Journal of Physiology*, **117**:165–181, 1952.
13. F.C. Hoppensteadt and E.M. Izhikevich. *Weakly Connected Neural Networks*. Springer-Verlag, New York, Berlin, and Heidelberg, 1997.
14. J.M. Hurtado, C.M. Gray, L.B. Tamas, and K.A. Sigvardt. Dynamics of tremor-related oscillations in the human globus pallidus: A single case study. *Proc. Natl. Acad. Sci. USA*, **96**:1674–1679, 1999.
15. E.M. Izhikevich. Neural excitability, spiking, and bursting. *International Journal of Bifurcation and Chaos*, **10**, 2000.
16. D. Johnston and S. M-S. Wu. *Foundations of Cellular Neurophysiology*. MIT Press, Cambridge, Ma. and London, 1997.
17. C.K.R.T. Jones. Stability of the traveling wave solutions of the fitzhugh-nagumo system. *Trans. Amer. Math. Soc.*, 286:431–469, 1984.
18. E.R. Kandel, J.H. Schwartz, and T.M. Jessell. *Principles of Neural Science*. Appleton & Lange, Norwalk, Conn., 1991.
19. J. Keener and J. Sneyd. *Mathematical Physiology*. Springer, New York, Berlin, Heidelberg, 1998.
20. N. Kopell and B. Ermentrout. Mechanisms of phase-locking and frequency control in pairs of coupled neural oscillators. In B. Fiedler, G. Iooss, and N. Kopell, editors, *Handbook of Dynamical Systems, vol. 3: Towards Applications*. Elsevier, 2003.
21. N. Kopell and D. Somers. Anti-phase solutions in relaxation oscillators coupled through excitatory synapses. *J. Math. Biol.*, **33**:261–280, 1995.
22. E. Lee and D. Terman. Uniqueness and stability of periodic bursting solutions. To appear in *Journal of Differential Equations*.
23. T. LoFaro and N. Kopell. Timing regulation in a network reduced from voltage-gated equations to a one-dimensional map. *J. Math. Biol.* To appear.
24. C. Morris and H. Lecar. Voltage oscillations in the barnacle giant muscle fiber. *Biophys. J.*, **35**:193–213, 1981.
25. M. Pernarowski. Fast subsystem bifurcations in a slowly varying Lienard system exhibiting bursting. *SIAM J. Appl. Math.*, **54**:814–832, 1994.
26. P. Pinsky and J. Rinzel. Intrinsic and network rhythmogenesis in a reduced Traub model for CA3 neurons. *J. Comput. Neurosci.*, **1**:39–60, 1994.
27. J. Rinzel. A formal classification of bursting mechanisms in excitable systems. In A.M. Gleason, editor, *Proceedings of the International Congress of Mathematicians*, pages 1578–1594. American Mathematical Society, Providence, RI, 1987.

28. J. Rinzel and G.B. Ermentrout. Analysis of neural excitability and oscillations. In C.Koch and I. Segev, editors, *Methods in Neuronal Modeling: From Ions to Networks*, pages 251–291. The MIT Press, Cambridge, MA, second edition, 1998.

29. J. Rinzel, D. Terman, X.-J. Wang, and B. Ermentrout. Propagating activity patterns in large-scale inhibitory neuronal networks. *Science*, **279**:1351–1355, 1998.

30. J. Rubin and D. Terman. Geometric analysis of population rhythms in synaptically coupled neuronal networks. *Neural Comput.*, **12**:597–645, 2000.

31. J. Rubin and D. Terman. Geometric singular perturbation analysis of neuronal dynamics. In B. Fiedler, G. Iooss, and N. Kopell, editors, *Handbook of Dynamical Systems, vol. 3: Towards Applications*. Elsevier, 2003.

32. A. Sherman. Contributions of modeling to understanding stimulus-secretion coupling in pancreatic β-cells. *American. J. Phyciology*, **271**:547–559, 1996.

33. P. Smolen, D. Terman, and J. Rinzel. Properties of a bursting model with two slow inhibitory variables. *SIAM J. Appl. Math.*, **53**:861–892, 1993.

34. D. Somers and N. Kopell. Rapid synchronization through fast threshold modulation. *Biol. Cybern.*, **68**:393–407, 1993.

35. D. Somers and N. Kopell. Waves and synchrony in networks of oscillators of relaxation and non- relaxation type. *Physica D*, **89**:169–183, 1995.

36. S.H. Strogatz. *Nonlinear Dynamics and Chaos*. Addison-Wesley Publishing Company, Reading, Ma., 1994.

37. J. Tabak, W. Senn, M.J. O'Donovan, and J. Rinzel. Comparison of two models for pattern generation based on synaptic depression. *Neurocomputing*, **26-27**:551–556, 1999.

38. D. Terman. Chaotic spikes arising from a model for bursting in excitable membranes. *SIAM J. Appl. Math.*, **51**:1418–1450, 1991.

39. D. Terman. The transition from bursting to continuous spiking in an excitable membrane model. *J. Nonlinear Sci.*, **2**:133–182, 1992.

40. D. Terman, N. Kopell, and A. Bose. Dynamics of two mutually coupled inhibitory neurons. *Physica D*, **117**:241–275, 1998.

41. D. Terman and E. Lee. Partial synchronization in a network of neural oscillators. *SIAM J. Appl. Math.*, **57**:252–293, 1997.

42. D. Terman, J. Rubin, A.C. Yew, and C.J. Wilson. Activity patterns in a model for the subthalamopallidal network of the basal ganglia. *J. Neuroscience*, **22**:2963–2976, 2002.

43. D. Terman and D. L. Wang. Global competition and local cooperation in a network of neural oscillators. *Physica D*, **81**:148–176, 1995.

44. D.H. Terman, G.B. Ermentrout, and A.C. Yew. Propagating activity patterns in thalamic neuronal networks. *SIAM J. Appl. Math.*, **61**:1578–1604, 2001.

45. X.-J. Wang and J. Rinzel. Spindle rhythmicity in the reticularis thalamic nucleus: synchronization among mutually inhibitory neurons. *Neuroscience*, **53**:899–904, 1993.

46. X.-J. Wang and J. Rinzel. Oscillatory and bursting properties of neurons. In M,A. Arbib, editor, *The Handbook of Brain Theory and Neural Networks*, pages 686–691. The MIT Press, Cambridge, London, 1995.

Neural Oscillators

Bard Ermentrout*

Department of Mathematics, University of Pittsburgh
Thackeray Hall 502, Pittsburgh, PA 15260, USA
bard@pitt.edu

1 Introduction

In this review, I will discuss a number of methods than can be used to mathematically analyze the behavior of coupled neural oscillators and networks of such units. This is by no means an exhaustive review and I refer the reader to the comprehensive reviews by Kopell and Ermentrout [39] or Rubin and Terman [55] which cover many more details. I will begin with a discussion of how oscillations arise in neural models. I then discuss the phase resetting curve, an experimental measure, and how this can be used to analyze coupled neural oscillators. A related method using discrete dynamics is applied to pairs of excitatory-inhibitory local networks. I then discuss a variety of general reduction procedures that lead to drastically simplified models suitable for use in large networks. I conclude with a brief discussion of network behavior.

Oscillations are observed throughout the nervous system at all levels from single cell to large networks. Some of the earliest experiments in physiology were aimed at understanding the underlying mechanisms of rhythmicity in nerve axons. Indeed, Hodgkin & Huxley won their Nobel Prize for dissecting the biophysical mechanisms underlying action potential generation in the giant axon of the squid. Oscillations are often associated with simple repetitive motor patterns such as walking, swimming, chewing, breathing, and copulation. Muscles responsible for these actions are controlled by the outputs of neurons whose outputs in turn are controlled by neural circuits in the brain and spinal cord. For this reason, they are called central pattern generators (CPGs). CPGs in the stomatogastric ganglion in the lobster have been well characterized and the exact wiring and neurons responsible for the activity are in some cases completely known [45]. CPGs produce a variety of regular rhythms that control the behavior of the target muscles. Thus, in the case of CPGs the importance of oscillatory behavior is clear and without doubt. Various models based on coupled oscillators have been suggested for quadruped lo-

---

* Supported in part by NIMH and NSF.

comotion ([6, 56, 14]). Undulatory swimming in the leech [4], the lamprey[13], and other animals is controlled by a series of segmental oscillators [12].

Oscillations in various areas of the brain are also associated with various pathologies. Epilepsy produces massive rhythmic or near rhythmic behavior throughout large areas of the brain. Parkinsonian tremors appear to be a consequence of pathological rhythmic behavior in an area of the brain called the basal ganglia. Large amplitude rhythmic electroencephalographic (EEG) activity is found in patients suffering from Creutzfeldt-Jacob disease. Oscillations occur throughout large regions of the brain during anaesthesia and sleep.

The possible role of oscillatory activity in normal sensory and cognitive behavior is more controversial. It has been suggested that synchronous oscillations in cortex and the phase and timing information they confer could be used to bind together different aspects of a sensory stimulus [31, 57]. Synchronous oscillations in the antennal lobe of insects appear to enhance the ability of the animal to distinguish between two closely related odors [58]. In [25], we suggested that synchronous oscillatory activity enhanced the competition different stimuli thus making it possible to direct attention the the more salient one. In a series of papers [20, 28] we studied the role of oscillations and waves in the learning of odors by the slug by modeling neurons in the procerebral lobe. Ulinski and coworkers [47] have analyzed stimulus-induced waves in the turtle visual area and suggest that the patterns of these waves are characterizing the stimulus in a global distributed fashion. Lam et al [42] suggest that odors are encoded by patterns of oscillations in the turtle olfactory bulb. In spite of the large experimental literature on these sensory-induced oscillations, there is still no definitive evidence for their role in cognition.

2 How Does Rhythmicity Arise

In this section, I discuss the mechanisms that generate periodic behavior from the point of view of nonlinear dynamics. I will only sketch the main points; a thorough and detailed account can be found in [54]. Suppose that we consider a nonlinear differential equation of the form:

$$\frac{dx}{dt} = F(x, \alpha) \tag{1}$$

where α is a parameter and $x \in R^n$. Most models of neurons have a resting state which corresponds to a fixed point of the system (1). As the parameter increases, we assume that a limit cycle arises. There are several mechanisms by which this can happen. The best known is the Hopf bifurcation (HB). In the HB, the fixed point $x_0(\alpha)$ becomes unstable at $\alpha = \alpha_0$ when a complex conjugate pair of eigenvalues of the linearized system cross the imaginary axis at $\pm i\omega$. Generically, a branch of periodic orbits emerges from this fixed point; these cycles have small amplitude and have the form:

$$x(t) = x_0(\alpha_0) + \epsilon z(t)\Phi e^{i\omega t} + cc$$

where "cc" means complex conjugates, Φ is a fixed complex vector depending on the problem and $z(t)$ is a complex scalar which satisfies

$$\frac{dz}{dt} = \epsilon[a(\alpha - \alpha_0)z + cz^2\bar{z}] \tag{2}$$

with a, c complex constants. If $\mathrm{Re}\,c < 0$ we say that the bifurcation is *super-critical* and the resulting branch of small amplitude periodic orbits is stable (Fig 1A). If $\mathrm{Re}\,c > 0$, then the bifurcation is *subcritical* and the resulting oscillations are unstable. In most neural models, the unstable branches turn around and produce stable periodic orbits which have large amplitude. This is shown in Figure 1B. This case is quite interesting since, for a range of parameters near α_0, there is both a stable fixed point and a stable large amplitude oscillation separated in phase space by the unstable small amplitude periodic orbit. The frequency of oscillations in a Hopf bifurcation is generally confined to a rather limited range (Fig 1D). Hodgkin [34] classified the firing patterns of nerve axons into two different classes : Class I and Class II excitability. (Also called Type I and Type II). In Rinzel and Ermentrout, we suggest that the properties of type II neurons are consistent with the dynamics resulting from a Hopf bifurcation.

A second common way for oscillations to emerge from a fixed point is through a saddle-node infinite-period bifurcation. Unlike the Hopf bifurcation, this is a global bifurcation and requires knowledge of the full phase space in order to prove. Figure 1C illustrates this bifurcation. There is a stable fixed point x_n and a saddlepoint x_s with a one-dimensional unstable manifold and an $n-1-dimensional$ stable manifold. The branches of the unstable manifold terminate on the stable fixed point forming a heteroclinic loop. As the parameter α changes, these two fixed points merge at a saddle-node bifurcation. However, the global loop still remains (see Figure 1C) and as the parameter passes through α_0, the fixed points disappear leaving a large amplitude stable periodic orbit. The frequency of this orbit can be arbitrarily low and scales as $\sqrt{\alpha - \alpha_0}$ (Fig 1F.) Rinzel and Ermentrout suggest that the properties of class I (Type I) neurons are consistent with this type of dynamics. The local behavior of a saddle-node is captured by the normal form [41]:

$$\frac{dx}{dt} = qx^2 + b(\alpha - \alpha_0) \tag{3}$$

where $q > 0, b > 0$ are problem dependent parameters. When $\alpha > \alpha_0$ (3) has no fixed points and any initial conditions tend to infinity in finite time. The real line can be mapped onto the circle so that this "blow-up" no longer occurs by making the change of variables, $x = \tan\theta/2$ leading to:

$$\frac{d\theta}{dt} = q(1 - \cos\theta) + b(1 + \cos\theta)[\alpha - \alpha_0]. \tag{4}$$

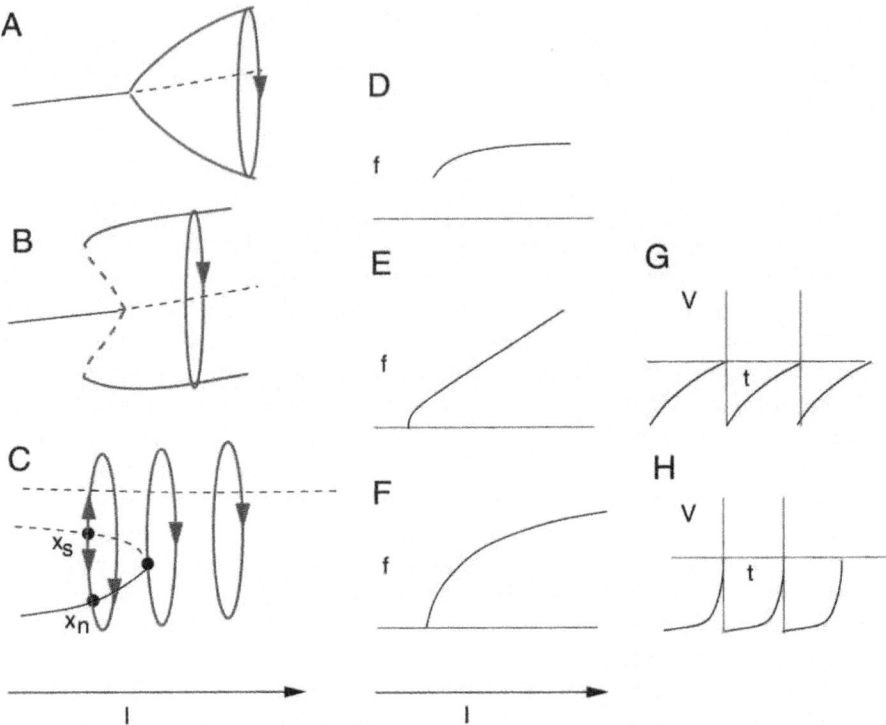

Fig. 1. Firing rates for neuron models. (A) Onset of periodicity through a supercritical Hopf bifurcation leading to small amplitude stable periodic orbits (B) Subcritical Hopf bifurcation showing large amplitude limit cycles and bistability between oscillations and the fixed point. (C) Saddle-node on a limit cycle bifurcation to large amplitude periodic orbits. (D) Frequency (firing rate) as a function of current for the bifurcations in A,B. (E) Frequency versus current for the leaky integrate-and-fire model (LIF). (F) Frequency versus current for the bifurcation C and for the quadratic integrate-and-fire model (QIF). (G) Voltage plots for the LIF model with spikes painted on. (H) Voltage plots for the QIF model.

In spite of the global nature of the dynamics, in [22] we show that the behavior near the saddle-node on a circle is captured by this simple scalar model. This model is often used to simulate spiking neurons and is called the "theta" model. It seems to be a better model for the dynamics of cortical inhibitory neurons than the Hopf bifurcation model. Note that under this change of variables, $\theta = \pi$ is equivalent to approaching $+\infty$ and "resetting" to $-\infty$ in the original equation (3). If, instead of letting x reach $+\infty$ in equation (3)and then resetting to $-\infty$ we reset to a finite value x_{reset} when x crosses x_{spike}, then we obtain a simple computational model for a spiking neuron called the quadratic integrate-and-fire (QIF) model. This was first introduced by Latham et al [43] in the following equivalent form:

$$\tau \frac{dV}{dt} = \frac{(V - V_{rest})(V - V_{thresh})}{V_{thresh} - V_{rest}} + R_m I \tag{5}$$

where I is the applied current, R_m is the membrane resistance of the cell, τ is a characteristic time constant, V_{rest} is the resting state of the neuron, and V_{thresh} is the threshold for excitability. Clearly as I increases, the two fixed points of this model merge at a saddle-node and for I large enough, disappear. The voltage V is reset to a finite value, $V_{reset} > -\infty$ when it reaches a finite value $V_{spike} < \infty$.

A simpler model for spiking neurons but one which cannot be derived from any type of bifurcation is the leaky integrate and fire model (LIF) which has the form:

$$\tau \frac{dV}{dt} = -(V - V_{rest}) + R_m I.$$

The LIF model is reset to V_{reset} whenever V crosses a value V_{spike}. The approach to a spike for the LIF model is concave down while that of the QIF is both. Figure 1E,F shows the output frequency of the LIF and the QIF models to a constant current, I. Note that due to its equivalence, the QIF and the saddle-node model have the same firing rate curve. Figures 1G,H show the shape of the membrane potential, $V(t)$ for several cycles; spikes are painted on whenever the spike-threshold is crossed. The LIF model is often used for analysis since it can be exactly solved even when I varies in time. This is not true for the QIF model and the theta model. (However, for sinusoidal stimuli, the theta model can be transformed into the Mathieu equation, [22]).

We point out one last model related to the QIF that is due to Izhikevich [36] and which can produce many of the spiking patterns of cortical neurons by changing parameters. If we add a simple linear negative feedback term to the QIF, we obtain the following model:

$$\frac{dV}{dt} = (v^2 + 125v)/25 + 140 + I - z \qquad \frac{dz}{dt} = a(bv - z)$$

along with the the reset conditions, when $v = +30$ then $v = c$ and $z = z + d$. Here I is the input and a, b, c, d are free parameters. Ermentrout and colleagues [17, 26] also considered this system when $b = 0$ as a model for high-threshold spike frequency adaptation. The additional term, b makes this similar to adding low-threshold adaptation.

3 Phase-Resetting and Coupling Through Maps

The stability properties of periodic orbits of autonomous differential equations are different from those of fixed points. There is always a zero eigenvalue which corresponds to translation in time of the orbit. This means that when a limit cycle is perturbed through an external stimulus, there can be a phase shift. This extra degree of freedom is what leads to the ability of several coupled oscillators to maintain differences in timing. It is also what is responsible for jet-lag; the internal circadian oscillator (body clock) needs to be rest by sunlight and this can take several days. Experimentalists have long known about the ability of oscillators to shift their phase and the quantification of this shift is one of the standard methods to characterize a biological oscillator. The basic thesis is that if I know how an oscillator shifts its phase to a stimulus, then I can use this to study entrainment by periodic stimuli and other oscillators. That is, given the *phase-resetting curve* (PRC) for an oscillator, it is possible to construct a map for the behavior when it is periodically driven or coupled to other similar oscillators.

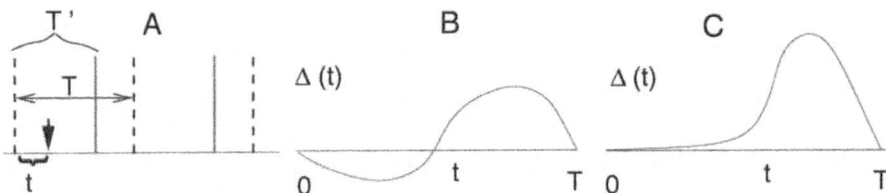

Fig. 2. Phase response curves (PRC). (A) Construction of the PRC. Natural period is T. A perturbation arrives at t after the last spike causing the spike to change its time to T'. The PRC, $\Delta(t) = 1 - T'/T$. Dashed pulses show the times at which the oscillator fires without the perturbatioin and the solid curves show the perturbed firing times. Note that a key assumption is that effect of the perturbation occurs only during the cycle it was given. (B) PRC for the firefly *P. Malaccae*. (C) PRC for a cortical neuron.

Figure 2A shows how the PRC is constructed. A convenient measure of the timing of a neuron is the time of its action potential, so we assume that the oscillator in question has a well define event corresponding to the start of a cycle. Suppose the period of the unperturbed oscillator is T and a brief stimulus is given at $t \in [0, T)$. This causes the oscillator to shift its timing (that is, the time of the next event) to a time T'. We define the PRC to be

$$\Delta(t) \equiv 1 - \frac{T'}{T}$$

If we define $\phi = t/T$ as the phase of the oscillator, we can define this function in terms of ϕ, $\Delta[\phi] = \Delta(\phi T)$. If $\Delta(t) > 0$ (respectively $\Delta(t) < 0$), this means that the oscillator fired early (late) and the phase was advanced (delayed). Certain species of SE Asian fireflies (notably, *Pteroptx mallacae* or *P. cribellata*) congregate by the thousands in trees along riverbanks and after some time, flash synchronously with a period of about a second. The the PRC for a single firefly has been measured [33, 5] to understand how such synchronization could take place. Figure 2B shows the form of the PRC for *P. mallacae* which is roughly sinusoidal. We note that the PRC for an oscillator near a Hopf bifurcation has essentially the same shape. Figure 2C shows the PRC form for a cortical neuron injected with constant current (to make it oscillate) and then subjected to brief depolarizing pulses [53]. The main difference between the two PRCs is that the cortical one is basically positive over the whole range of perturbation times. Positive PRCs are characteristic of model neurons which become periodic through a saddle-node bifurcation (Type I). Suppose that we parameterize the PRC by the amplitude of the perturbing stimulus, $\Delta(t, a)$. Then consider

$$Z(t) \equiv \lim_{a \to 0} \frac{\Delta(t, a)}{a}.$$

This is called the infinitesimal PRC. Now suppose the model for the oscillator has the form:

$$\frac{dV}{dt} = F(V, w_1, \ldots, w_n), \quad \frac{dw_j}{dt} = g_j(V, w_1, \ldots, w_n)$$

and the perturbation is applied only to the equation for V. Let $(V^0(t), w_j^0(t))$ be the limit cycle solution to this equation with the maximum of V occurring at $t = 0$. Then, the infinitesimal PRC, $Z(t)$ is proportional to the V component of the adjoint solution (see [32, 23]). Recall that if $X' = F(X)$ has a periodic solution, $X_0(t)$, then the adjoint, $X^*(t)$ satisfies $X^{*\prime}(t) = -D_X F(X_0(t))^T X^*(t)$ with $X^{*T}(0)X_0'(0) = 1$. This makes the calculation of the PRC for models simple, one need only compute the adjoint to a given periodic solution. This also allows one to explicitly calculate the PRC for certain simple oscillators. The PRC for the theta model, equation (4), is proportional to $(1 - \cos(t))$ while that for a supercritical Hopf bifurcation, equation (2), is proportional to $\sin(t)$ where the period of the oscillation is 2π and the spike occurs at $t = 0$. Many PRCs in biology satisfy $\Delta(0) = \Delta(T) = 0$.

Looking at the PRC for the firefly (figure 2B), it is intuitively easy to see how a pair of mutually coupled oscillators could synchronize. Suppose oscillator 1 fires slightly before oscillator 2. Then it will speed up oscillator 2 since the PRC is positive right before the spike. Once oscillator 2 fires, this will slow down oscillator 1 since it has recently fired. Thus, the oscillator ahead is slowed down and the one behind is sped up. We can formalize by creating a model for a pair of mutually coupled oscillators. But before doing this, we first consider an oscillator that is periodically driven by a small pulsatile stimulus with a period P relative to the period of the oscillator. Let ϕ_n be the phase of an oscillator ($\phi \in (0, 1]$) at the point the nth stimulus arrives. The phase is incremented by an amount $\Delta(x_n)$ immediately after the stimulus. Then between stimuli, the oscillator advances by an amount P, so that we obtain

$$\phi_{n+1} = \phi_n + \Delta(\phi_n) + P \equiv G(\phi_n; P) \tag{6}$$

We will assume that the stimulus is weak enough so that the function $F(x) = x + \Delta(x)$ is monotone increasing. (This means that $\Delta'(t) > -1$.) We also assume that $\Delta(0) = \Delta(1) = 0$. This implies that the map $F : [0, 1) \longrightarrow [0, 1)$

is invertible. This has many implications for the forced oscillator; notably that the ratio,

$$\rho = \lim_{n \to \infty} \frac{\phi_n}{n}$$

is almost everywhere rational and a continuous function of the parameters of the model. It is called *rotation number*. On regions where ρ is rational, we say that the oscillator is phaselocked to the stimulus. One solution of interest to (6) is 1:1 locking. This solution requires that $\rho = 1$ and that $\phi_{n+1} = \phi_n + 1$. The latter definition implies that $\phi_n = \bar{\phi}$ satisfies

$$1 - P = \Delta(\bar{\phi}).$$

Since $\Delta(\phi)$ is periodic, solutions are lost in pairs at saddle-node bifurcations. (For example, suppose that $\Delta(\phi) = -a \sin 2\pi\phi$. Then if $|1 - P| > |a|$ there are no fixed points and if $|1 - P| < |a|$ there are two fixed points; a saddle-node occurs at $|1 - P| = |a|$.) For a PRC such as the firefly's, it is possible to obtain 1:1 locking for periods, P both shorter and longer than the natural period of the oscillator. However, for oscillators with a PRC like type I oscillators or like the cortical neuron, it is only possible to lock to stimuli that are faster than the intrinsic period, *i.e.*, $P < 1$. This is intuitively obvious; stimuli can only speed up the oscillator since the PRC is positive. A 1:1 locked solution is stable if and only if

$$|F'(\bar{\phi})| = |1 + \Delta'(\bar{\phi})| < 1.$$

By hypothesis, $\Delta'(\phi) > -1$ so that the necessary condition for stability is that $\Delta'(\bar{\phi}) < 0$. For the PRCs illustrated above, the solution stable solution occurs on the falling side of the PRC.

We now turn to the analysis of the behavior of a pair of identical reciprocally coupled oscillators ([30]). We suppose that the period of each is 1 without loss in generality. We will write the coupled system, formally, as a set of differential equations and from this derive a locally defined map near a particular periodic solution:

$$\dot{\theta}_1 = 1 + \delta(\theta_2)\Delta(\theta_1), \quad \dot{\theta}_2 = 1 + \delta(\theta_1)\Delta(\theta_2). \tag{7}$$

Each time, say, oscillator 1 fires (that is, crosses 0 modulo 1), then oscillator 2 has its phase altered as $\theta_2 = \theta_2 + \Delta(\theta_2) \equiv F(\theta_2)$. The function, $F(\theta)$ is called the phase-transition curve. Note that our assumptions on $\Delta(\theta)$, namely that $\Delta'(\theta) > -1$, and that $\Delta(0) = \Delta(1) = 0$ imply that $F(\theta)$ is monotone, with $F(0) = 0$ and $F(1) = 1$. Suppose that θ_1 fires a spike at $t = 0$ and $\theta_2 = \phi$. Then, immediately after the spike, $\theta_2 = F(\phi) < 1$ since F is monotone and $F(1) = 1$. θ_2 will spike at $t_2 = 1 - F(\phi)$ at which point θ_1 has advanced by an amount t_2. The spike elicited by θ_2 advances θ_1 to $F(1 - F(\phi))$ while θ_2 is set to 0. θ_1 will next fire at $t_1 = 1 - F(1 - F(\phi))$ and θ_2 has advanced by the same amount. Thus, we have the map

$$\phi \longrightarrow 1 - F(1 - F(\phi)) \equiv Q(\phi).$$

Note that $Q(0) = 0$ so that $\phi = 0$ is always a fixed point of the map. Other fixed points are found by solving $\phi = Q(\phi)$. The fixed point, $\bar{\phi}$ is stable if $|Q'(\bar{\phi})| < 1$ which translates to $F'(1 - F(\bar{\phi})) \cdot F'(\bar{\phi}) < 1$ since $F'(\phi) > 0$. This is the product of two terms; the derivative of the phase-transition curve evaluated at the phase of each oscillator when the other fires a spike. In particular, consider the synchronous solution. This will be stable if $F'(1^-)F'(0^+) < 1$ or, $[1 + \Delta'(1^-)][1 + \Delta'(0^+)] < 1$. Consider the firefly PRC. Then since $\Delta'(0^+) < 0$ and $\Delta'(1^-) < 0$, this means that synchrony is always stable. On the other hand, it is not necessary that the derivative of Δ is continuous at 0, so that it may be that case that $\Delta'(1^-) = \Delta'(0^-) \neq \Delta'(0^+)$. This appears to be the case for cortical PRCs for which $\Delta'(0^+) > 0$ and $\Delta'(1^-) < 0$. For the PRC shown in figure 2C, for example, since the slope at 0 is positive but shallow and the slope at 1 is steep and negative, this implies that $Q'(0) < 1$ so that synchrony is stable.

We can use these general methods to create networks of coupled cells through their PRCs:

$$\dot{\theta}_j = \omega_j + \sum_k \delta(\theta_k) \Delta_{jk}(\theta_j),$$

where we allow for some slight differences in natural frequencies, ω_j. In [30], we consider rings of cells for which we prove the existence and stability of traveling waves. We also consider two-dimensional arrays of cells with nearest neighbor interactions. We show the existence of rotating waves in such a network.

We close this section with a simplification of the general pulse-coupled network:

$$\frac{d\theta_j}{dt} = \omega + \sum_k g_{jk} P(\theta_k) \Delta(\theta_j). \tag{8}$$

In this model, the Dirac function is replaced by a smooth pulse function and all cells have the same PRC; the only difference lies in the coupling amplitudes, g_{jk}. This model was suggested by Winfree [64]. Suppose that $g_{jk} \geq 0$ and that $G = \sum_k g_{jk}$ is independent of j. Let

$$\frac{d\theta}{dt} = \omega + GP(\theta)\Delta(\theta)$$

have a solution $\theta(t) = \phi(t)$ with $\phi'(t) > 0$ and $\phi(0) = 0, \phi(T) = 1$ for some $T > 0$. Then (i) there is a synchronous solution to (8), $\theta_j(t) = \phi(t)$ and (ii) it is asymptotically stable if

$$\int_0^T P(\phi(t)) \Delta'(\phi(t)) \, dt < 0.$$

We remark that this depends crucially on the non-negativity of g_{jk}. We also note that if $\Delta(\theta)$ is similar to the firefly in that it is decreasing around the origin and $P(\theta)$ is sharply peaked around the origin, nonnegative, and vanishes outside some small neighborhood of the origin, then the stability condition will hold.

4 Doublets, Delays, and More Maps

In [24] we used maps to explain the importance of inhibitory doublets for synchronizing two distant populations of neurons that were producing a so-called gamma rhythm of about 40 Hz. Each population consists of excitatory and inhibitory neurons which are locally coupled in such a way that they are synchronous within the population. In the local populations, there are few (or weak) excitatory-excitatory connections. Between distant groups, the excitatory cells of one group connect to the inhibitory cells of the other group and because the groups are far apart, the connection between them has a delay. In large scale simulations of this setup [59, 60], thye found that the inhibitory population of cells fires a pair of spikes (called a doublet) when the two populations of excitatory cells were synchronized. Our approach to this problem was to consider a firing time map for the spike times of the excitatory population. The basic ideas are as follows. E cells cannot fire until the inhibition from them has worn off below a certain level. Each time an I cell fires the synapse to the E cell is set to its maximum conductance and then it decays exponentially. When an I cell receives excitation, it fires a spike at a time that depends on the last time that it spiked. In particular, if the I cell has recently spiked, then an incoming E spike may not cause the I cell to spike (absolute refractoriness) or if it does spike, it is with some perhaps lengthy delay. Indeed, because the I cell models are class I excitable, they can fire with an arbitrarily long latency if the stimulus is timed right. We now derive a simple map for the synchronization of these two groups of oscillators. The key feature that is responsible for the synchronization of the pair is the recovery map for the inhibitory neurons, $T_I(t)$. Specifically, $T_I(t)$ is the time that the I cell fires its next spike given that it has received an E spike t milliseconds since it last fired. This map can be computed numerically by firing an E cell at different times after the I cells has spiked and measuring the latency to the next spike. For t large, the map approaches a small positive latency, T_{ei}. For t small, the map is rather steep and in fact, it may be undefined for t small enough. Recurrent inhibitory synapses make the map steeper since they act to make the I cell even more refractory. Stronger excitatory inputs make the map shallower and lower the latency for all times. The map can be approximated by the function:

$$T_I(t) = T_{ei} + \frac{a}{t+b}.$$

If $b < 0$, then the map is undefined for $t < -b$.

A brief digression Allow me to digress briefly and derive this map from the biophysics of the I cell. The I cell is a type I neuron so that at the bifurcation, the dynamics is approximated by $dx/dt = qx^2$ (see equation (3). Suppose that the impulse from the E cell arrives at t^* and $x(0) = -x_0$ is the degree of refractoriness of the I cell right after firing. Until the I cell receives input,

$$x(t) = \frac{-1}{1/x_0 + qt}$$

and when the input comes

$$x(t^{*+}) = A - \frac{1}{1/x_0 + qt^*}$$

where A is the amplitude of the input. Firing is defined as when the solution "blows up" which occurs when

$$t_{fire} \equiv \frac{q}{A - \frac{1}{1/x_0 + qt^*}}$$
$$= T_{ei} + \frac{a}{t^* + b}$$

where

$$T_{ei} = \frac{q}{A}, \quad a = A^{-2}, \quad \text{and } b = \frac{1}{q}\left(\frac{1}{x_0} - \frac{1}{A}\right).$$

With this little derivation, we can assign meaning to each of the parameters. The larger the refractory period, x_0, the more negative the parameter b. Similarly, the larger the value of the excitatory stimulus, the shallower the map. For sufficiently large stimulus, the map is defined for $t = 0$, i.e., $T_I(\theta) > 0$.

End of digression

We assume that there is a conduction delay, δ, of the impulse from an E cell in one group to the I cell in the other group. We are interested in the ability of the two groups to synchronize stably in the presence of this delay. We suppose that an E cell fires a fixed amount of time T_{ie} after the last I spike that it has received in a cycle. This is not unreasonable given our assumption that the I cell synapses always produce their maximal conductance when the I cells spike. This also implicitly assumes that the E cells have no slow processes that are turned on when the E cell spikes. (For example, strong spike-frequency adaptation would make the present analysis suspect. Indeed in [40] we include adaptation in the excitatory cells and show that the behavior is quite different. In [24], we in fact incorporate some memory of the previous spikes for the E cells, but to keep matters simple, we ignore it here.) Let t_j be the time of firing of the excitatory neuron in group j ($j = 1, 2$) at the start of a cycle. By this time the I cells in that group are completely recovered so that they fire at a time, $t_j + T_{ei}$. The I cell in group 1 receives an extrinsic E spike at time $t_2 + \delta$, so that this I cell will fire another spike (the "doublet") at a time,

$$t_1^{doub} = t_2 + \delta + T_I[t_2 + \delta - (t_1 + T_{ei})]$$

The E cell in group 1 will then fire its next spike at $t_1^{new} = t_1^{doub} + T_{ie}$. This leads to the following map:

$$t_1^{new} = t_2 + \delta + T_I(t_2 - t_1 + \delta - T_{ei}) + T_{ie}$$
$$t_2^{new} = t_1 + \delta + T_I(t_1 - t_2 + \delta - T_{ei}) + T_{ie}.$$

We let ϕ be the timing difference between the two groups, $\phi = t_2 - t_2$ so that ϕ satisfies

$$\phi^{new} = -\phi + T_I(-\phi + \delta - T_{ei}) - T_I(\phi + \delta - T_{ei}) \equiv G(\phi). \qquad (9)$$

Clearly, $G(0) = 0$ so that synchrony (0 time difference) is a solution. Our interest is in its stability. This requires that $|G'(0)| < 1$ or $|-1-2T_I'(\delta - T_{ei})| < 1$ which simplifies to $T_I'(\delta - T_{ei}) > -1$ since T_I' is always negative. Applying this to our approximate map, we see that we must have

$$\delta > a - b + T_{ei} = A^2 + \frac{q}{A} + \frac{1}{q}\left(\frac{1}{A} - \frac{1}{x_0}\right).$$

However, for large values of δ, T_I' is nearly zero so that $G'(0)$ is close to -1 so that small heterogeneities in the two networks (say, T_{ie} is different from one to the other) will lead to an inability to lock, not through stability, but through existence. That is, if, for example, the two oscillators have different values of T_{ie}, then equation (9) is replaced by

$$\phi^{new} = G(\phi) - B$$

where B is the difference between the values of T_{ie}. Fixed points satisfy:

$$T_I(-\phi + \delta - T_{ei}) - T_I(\phi + \delta - T_{ei}) = B.$$

For large δ the left-hand side is nearly zero so that if B, the heterogeneity, is too large, there are no fixed points.

The maps derived in this section allow us to make some general conclusions about the roles of different parameters. First, if the delay is too small, then the slope of the T_I map is large and negative so that we expect to see instabilities. Increasing the $I - I$ coupling makes the map steeper since it essentially makes the I cells more refractory. The steeper the T_I map, the more difficult it is to stably lock for short delays. On the other hand, a steep T_I map is better able to overcome intrinsic differences between the two networks at long delays. Conversely, increasing the $E - I$ coupling makes the network less susceptible to instabilities for short delays but less able to compensate for heterogeneity.

5 Averaging and Phase Models

One of the assumptions implicit in the use of phase-resetting curves is that the coupling is not too strong. We can be more explicit about this "weak" coupling assumption by applying averaging to coupled oscillators. Consider the following coupled system in $R^n \times R^n$

$$X_1' = F(X_1) + \epsilon G_1(X_2, X_1), \quad X_2' = F(X_2) + \epsilon G_2(X_2, X_1). \tag{10}$$

We assume that when $\epsilon = 0$, the system $X' = F(X)$ has a phase-asymptotically stable limit cycle solution, $X_0(t)$ with period, T. The limit cycle is homeomorphic to the circle, S^1. Stability means that there is an open neighborhood, N of S^1 in R^n such that all point in N are attracted to the limit cycle. Thus, when $\epsilon = 0$, there is an attracting two-torus, $S^1 \times S^1$ and all points $(X_1, X_2) \in N \times N$ are attracted to the torus. Each point on the stable limit cycle can be parametrized by a single variable, θ, so that when $\epsilon = 0$, the dynamics on the torus have the form

$$\theta_1' = 1 \qquad \text{and} \quad \theta_2' = 1.$$

The solutions to these uncoupled equations are $\theta_j = t + \phi_j$ where ϕ_j are arbitrary constants. That is, without coupling, the solutions can have arbitrary phases as they move around the torus. For $\epsilon \neq 0$ and sufficiently small we expect that the torus will persist, but that the parallel flow on the torus will collapse toward a phaselocked solution where the difference, $\theta_2 - \theta_1$ takes on isolated values. Indeed, we expect that for nonzero ϵ the equations have the form $\theta_j' = 1 + \epsilon h_j(\theta_j, \theta_k)$. Our goal in this section is to discern the form fo h_j.

In order to derive the equations for $\epsilon > 0$, we introduce some notation. Let B denote the space in R^n of continuous T-periodic solutions. We introduce the inner product

$$< u(t), v(t) > = \frac{1}{T} \int_0^T u(t) \cdot v(t) \; dt.$$

$X_0(t)$ is the asymptotically stable limit cycle solution satisfying

$$X_0'(t) = F(X_0(t))$$

so that differentiating this, we see that the linear operator

$$(LY)(t) \equiv -\frac{dY(t)}{dt} + D_X F(X_0(t))Y(t) \equiv [-\frac{d}{dt} + A(t)]Y(t)$$

has a one-dimensional nullspace spanned by $\frac{dX_0(t)}{dt}$. Under the above innerproduct, the adjoint operator is

$$(L^* Z)(t) = [\frac{d}{dt} + A^T(t)]Z(t)$$

and this has a one-dimensional nullspace spanned by $Z^*(t)$ which can be normalized to satisfy $Z^*(t) \cdot X_0'(t) = 1$. (For asymptotically stable limit cycles, we can numerically compute $Z^*(t)$ by integrating $(L^* u)(t)$ backwards in time until convergence to a periodic solution is obtained.) Recall from section 3 that $Z^*(t)$ is related to the infinitesimal phase-response curve (see also below). Using the method of averaging [23] it is easy to derive the equations for the evolution of phases, $X_j(t) = X_0(\theta_j) + O(\epsilon)$ where

$$\theta_1' = 1 + \epsilon H_1(\theta_2 - \theta_1) + O(\epsilon^2) \quad \text{and} \quad \theta_2' = 1 + \epsilon H_2(\theta_1 - \theta_2) + O(\epsilon^2)$$

with

$$H_j(\phi) = \frac{1}{T} \int_0^T Z^*(t) \cdot G_j(X_0(t + \phi), X_0(t)) \, dt.$$

Before turning to the analysis of this equation, we apply the formalism to coupled neurons. Single compartment models are generally coupled through the potentials, $V(t)$. Suppose the equations for the voltage have the form:

$$\frac{dV_j}{dt} = \frac{1}{C} \left(-I_{ion}^j(t) + I_{syn}^{jk}(t) \right),$$

where $I_{ion}^j(t)$ are all the ionic currents intrinsic to the cell and $I_{syn}^{jk}(t)$ has the form

$$I_{syn}^{jk}(t) = g_{syn}^{jk} s_k(t)(V_j(t) - E_{syn}^{jk}) + g_{gap}^{jk}(V_j(t) - V_k(t)).$$

The variable $s_k(t)$ is gating variable just like the usual voltage-gated ion channels, but it is gated by the presynaptic voltage, V_k rather than the postsynaptic voltage. The second coupling term is called electrical coupling and depends on the voltage difference between the pre- and post-synaptic cells. Let $V^*(t)$ be the voltage component of the adjoint, $Z^*(t)$; recall that this is proportional to the PRC for the neuron, at least to lowest order. Then

$$H_j(\phi) = \frac{1}{C} \frac{1}{T} \int_0^T V^*(t)$$
$$\times \left(g_{syn}^{jk} s_0(t + \phi)(E_{syn}^{jk} - V_0(t)) + g_{gap}^{jk}(V_0(t + \phi) - V_0(t)) \right) \, dt.$$

Here $V_0(t), s_0(t)$ are the solutions to the uncoupled system on the limit cycle. Typically, $s_0(t)$ is qualitatively like, $a^2 t e^{-at}$ where t is the time since the neuron spiked during the cycle. Consider the first term in the integral:

$$\frac{1}{C} \frac{1}{T} \int_0^T V^*(t)(E_{syn}^{jk} - V_0(t)) s_0(t + \phi) \, dt \equiv H_c^{jk}(\phi)$$

The terms multiplying s_0 depend only on the postsynaptic cell (the one receiving the synapse). This term characterizes the response of a neuron to the synapse. Thus the first term in the integral is the average of the response with a phase-shift of the input. Note that in general, $H_c(0) \neq 0$. The second term in the integral is

$$\frac{1}{C} \frac{1}{T} \int_0^T Z^*(t)(V_0(t + \phi) - V_0(t)) \, dt \equiv H_g(\phi). \tag{11}$$

Note that $H_g(0) = 0$. We make one more observation. Suppose that the coupling from one oscillator to the other is delayed by an amount, say, τ. Then the only effect on the interaction function H is to shift the argument by τ. Intuitively, this is because θ is essentially a time variable so that delays are just translations in time.

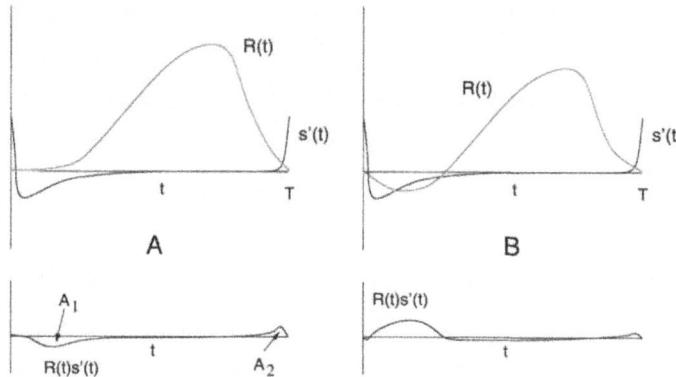

Fig. 3. Computation of the stability of synchrony. (A) the derivative of the synapse, $s_0'(t)$ and the response function, $R(t)$. The integral of their product is $H'(0)$ and is negative. (B) If $R(t)$ has a negative region after the spike, the area is positive and synchrony is stable.

Suppose that the two oscillators are coupled identically. Then we can write

$$\theta_1' = 1 + \epsilon H(\theta_2 - \theta_1), \quad \theta_2' = 1 + \epsilon H(\theta_1 - \theta_2),$$

where

$$H(\phi) = g_{gap}H_g(\phi) + g_{syn}H_c(\phi)$$
$$H_g(\phi) = \frac{1}{CT} \int_0^T Z^*(t)(V_0(t + \phi) - V_0(t))\, dt,$$
$$H_c(\phi) = \frac{1}{CT} \int_0^T Z^*(t)s_0(t + \phi)(E - V_0(t))\, dt.$$

Let $\phi = \theta - 2 - \theta_1$. Then

$$\phi' = \epsilon H(-\phi) - H(\phi).$$

The right-hand side is a continuous odd-periodic function so that it has at least two roots, $\phi = 0$ and $\phi = T/2$. The former represents the synchronous solution. It is a stable solution if and only if $-2H'(0) < 0$ which means that $H'(0) > 0$. For chemical synapses,

$$H'(0) = \frac{1}{T} \int_0^T R(t)s_0'(t)\, dt,$$

where $R(t) = Z^*(t)(E - V_0(t))$. Consider first excitatory coupling. For cortical neurons, $R(t)$ has the shape shown in figure 2C. Figure 3 shows $s_0'(t)$, $R(t)$ and their product. From this figure it is also clear that the area of the product $R(t)s_0'(t)$ is negative so synchrony is unstable. Less clear, but shown in specific examples is the fact that the anti-phase solution, $\phi = T/2$ is stable for excitatory coupling. For a PRC like that of the firefly which has a substantial negative region right after the spike (see figure 2B), synchrony is stable as seen in Figure 3B. Since inhibitory coupling essentially reverses the sign of $E - V_0(t)$, synchrony is stable and the anti-phase state is unstable [61, 29, 11, 63]. The behavior of gap junctions is more subtle and depends strongly on the actual shape of the action potential [10]; from the above formula:

$$H_g'(0) = \frac{1}{CT} \int_0^T Z^*(t)V_0'(t) \, dt.$$

Suppose that $Z^*(t)$ is positive and essentially zero and the action potential is short-lasting. Then $V_0'(t)$ is negative only at the beginning of the cycle where $Z^*(t)$ is small. For the majority of the cycle, $V_0'(t) > 0$ as the neuron repolarizes, so that the area under the integral is positive. Thus, we expect that for thin action potentials, synchrony is stable. However, as the frequency of the neuron increases, the action potential occupies a greater portion of the cycle and it may be possible to destabilize synchrony. These simple intuitive arguments are no replacement for the actual calculations. However, they serve to shed insight into how different forms of coupling alter the locked states and how these states depend on frequency.

5.1 Local Arrays

So far, we have considered pairs of oscillators and their phaselocked behavior. I will next consider one-dimensional geometries and then use results from this to look at planar arrays of oscillators. Since rings are easier to deal with (there are no boundary effects), we start with them and illustrate traveling waves and synchrony. We then turn our attention to linear arrays in which the edge conditions play a pivotal role it determining the steady state patterns.

The ideas discussed above can be extended in an obvious fashion to large networks of weakly coupled nearly identical cells. This leads to equations of the form:

$$\theta_j' = \omega_j + H_j(\theta_1 - \theta_j, \dots, \theta_N - \theta_j), \quad j = 1, \dots, N. \tag{12}$$

A *phaselocked state* is one for which $\theta_j(t) = \Omega t + \phi_j$ where $\phi_1 = 0$ and the remaining ϕ_j are constants. Suppose that there is such a state and let

$$a_{jk} = \left. \frac{\partial H_j(\psi_i, \dots, \psi_N)}{\partial \psi_k} \right|_{\phi_1 - \phi_j, \dots, \phi_N - \phi_j}.$$

With these preliminaries, we can now state a very general stability theorem.

Theorem 1. [16] *Suppose that $a_{jk} \geq 0$ and the matrix $A = (a_{jk})$ is irreducible. Then the phaselocked state is asymptotically stable.*

This provides a quick sufficient check for stability. If we regard the oscillators as nodes and draw directed line segments from j to k if $a_{jk} > 0$, then irreducibility of the matrix A says that we can go from any given node i to any other node, ℓ by following these directed segments. We also note that stability implies that there are no zero eigenvalues of A other than the simple one corresponding to the translation invariance of limit cycles. Thus, the theorem also provides conditions under which we can continue solutions using the implicit function theorem as some parameter varies. We will apply some of these principles below.

5.1.1 Rings

Since a general ring of oscillators (even phase models) is impossible to completely analyze, we will assume translation invariance in the model system. That is, we assume the coupling between oscillators depends only on the differences between their indices. Furthermore, we assume the nature of the coupling between any two oscillators is the same and only the coefficient varies. Thus, we restrict our attention to the system

$$\theta'_j = \omega + \sum_{k=1}^{N} C(j - k)H(\theta_j - \theta_k) \tag{13}$$

where as usual, H is a sufficiently differentiable $2\pi-$periodic function. Equation (13) admits a family of traveling wave solutions of the form

$$\theta_j = \Omega_m t + 2\pi j m/M, \quad m = 0, \dots N - 1.$$

Substitution of this form into (13) implies that

$$\Omega_m = \omega + \sum_k C(k)H(2\pi km/N).$$

This provides the so-called dispersion relation for the waves; that is, how the frequency depends on the wave number, m. Stability of these wave is readily found by linearizing about the solution. The linearized system satisfies

$$y'_j = \sum_k C(j - k)H'(2\pi(j - k)m/N)(y_j - y_k).$$

Because the resulting linearization matrix is circulant, the eigenvectors are of the form $\exp(2\pi i j\ell/N)$ so that for a given m, we have the N eigenvalues

$$\lambda_{\ell,m} = \sum_k C(k)H'(2\pi km/N)(1 - \exp(-2\pi ik\ell/N)).$$

In particular, the synchronous solution has $m = 0$ and the eigenvalues are

$$\lambda_{\ell,0} = H'(0) \sum_k C(k)(1 - \exp(-2\pi i k \ell / N)).$$

If $H'(0) < 0$ and $C(k) \geq 0$, then it follows from Theorem 1 that synchrony is stable. However, if C changes sign, then synchrony is often unstable as are waves with a high value of m, but for some range of values of m, there are stable waves. As an example, consider a network of 100 oscillators with coupling between the 10 neighbors on either side, $C(j) = 1/9$ for $|j| < 5$ and $C(j) = -1/12$ for $10 \geq |j| \geq 5$. The function $H(x) = 0.5 \cos x - \sin x$. Then, synchrony is unstable, but the wave with $m = 4$ is stable. Random initial data for this ring network tend to a solution which is a mix of waves propagating in both directions.

5.1.2 Linear Arrays

Linear arrays of oscillators, even with only nearest neighbor coupling, are considerably more difficult to analyze. Unless the boundary conditions are chosen very specially, the general behavior is nontrivial. However, with sufficiently long chains, it turns out that the behavior is captured by the solutions to a certain two-point singularly perturbed boundary value problem. Consider the following network:

$$\theta'_j = \omega_j + H_f(\theta_{j+1} - \theta_j) + H_b(\theta_{j-1} - \theta_j), \quad j = 1, \ldots, N, \qquad (14)$$

where at $j = 1$ ($j = N$) we drop the H_b (H_f) term. In [37, 38], we showed that the phaselocked behavior of this system satisfies

$$\theta_{j+1} - \theta_j \longrightarrow \Phi(j/N)$$

where

$$\Omega = \omega(x) + H_f(\Phi) + H_b(-\Phi) + \frac{1}{N}[H'_f(\Phi) - H'_b(-\Phi)]\Phi_x \quad 0 < x < 1$$
$$\Omega = \omega(0) + H_f(\Phi(0))$$
$$\Omega = \omega(1) + H_b(-\Phi(1)).$$

For large N, this is a singularly perturbed first order differential equation. As with many singularly perturbed systems, there are boundary layers. In these papers, we exhaustively classify all possible phaselocked solutions for N large by studying the singular boundary value problem and through the use of matching of inner and outer equations. In figure 4, we illustrate various solutions to equation (14) for $N = 80$. Note that for the isotropic case (panel E), the boundary layer is in the interior. Note also, the value of N for which the continuum approximation is reasonable, depends on the parameters of

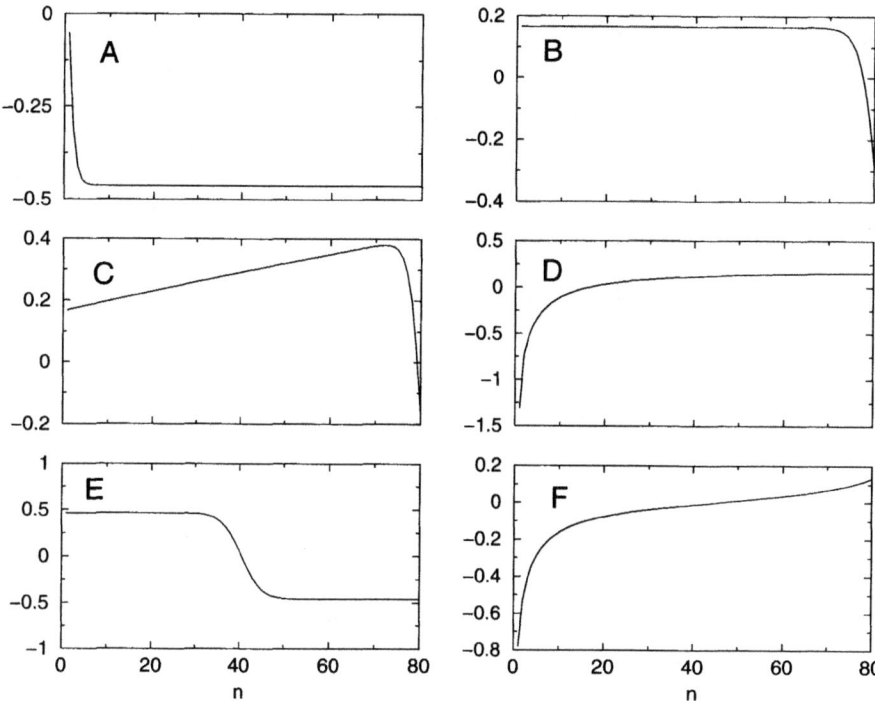

Fig. 4. Local phase differences in a chain of 80 nearest-neighbor oscillators under a variety of cases. (A,B,D,F) anisotropic coupling; (C) isotropic coupling with a linear gradient in frequency; (E) isotropic coupling.

the model. For example, panel F shows little evidence of a boundary layer although one is predicted at both sides for N large enough.

As another example of the behavior of one-dimensional chains, we consider the completely isotropic case with nearest neighnbors in which $H_f(\phi) = H_b(\phi) = H(\phi)$. For large N, the boundary layer (if it exists) will be interior. Suppose that $H(\phi) = C + g(\phi)$ where g is an odd periodic function like $\sin \phi$. Then, $\Phi(x) \approx K(x - 1/2)$ so that, the gradient is linear. In the second example, $H(\phi) = g(\phi + C)$ where, again, g is an odd periodic function. In this case, if g is similar to sin and $|C| < \pi/4$, then $\Phi(x) \approx A\text{sign}(x - 1/2)$. In both of these examples, the phase difference switches signs in the middle so that physically, one observes either waves propagating from the ends toward the middle or vice versa. These examples will play a role shortly.

5.1.3 Frequency Plateaus

In [21], we considered the behavior of a chain of oscillators with a linear frequency gradient. The motivation for this comes from experiments on waves in the small bowel. Smooth muscle in the bowel generates rhythmic oscillations. A plot of the frequency as a function of distance from the stomach (towards the large bowel) shows that over large regions, the frequency is constant, but, there are abrupt jumps to lower frequencies. When small pieces of the bowel are isolated, they show a linear gradient in natrural frequencies. Figure 5 shows an schematic of the frequency of the intrinsic and the coupled oscillators. We were motivated to study the behavior of a chain of locally coupled oscillators with a linear gradient in frequency:

$$\theta'_j = \omega - gj + H(\theta_{j+1} - \theta_j) + H(\theta_{j-1} - \theta_j) \quad j = 1, \ldots, N+1.$$

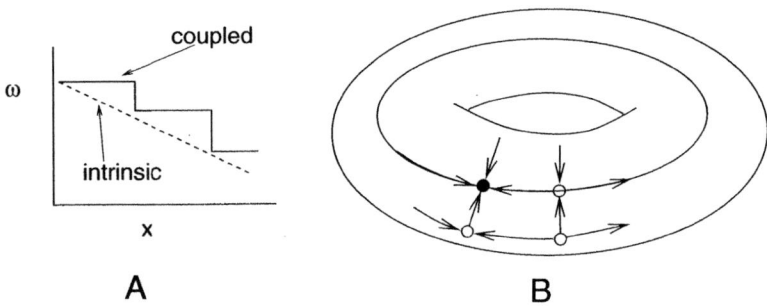

A   **B**

Fig. 5. Frequency plateaus in coupled oscillators. Left panel shows the intrinsic linear gradient in frequency and the coupled frequencies. Right panel gives a geometric picture of how plateaus are formed.

Since coupling in the bowel is through gap junctions, $H(0) = 0$. (See equation (11). Suppose, in addition, that $H'(0) > 0$. Then, for $g = 0$, $\theta_j = \Omega t + C$ is a solution. From Theorem 1, it is a stable solution and so we can continue the phaselocked state for g small. As g increases (that is, the frequency gradient increases), phaselocking can become disrupted. Thus, we can ask what happens for large g. This question can be answered to some satisfaction if we make the simplification that the function $H(\theta)$ is odd. For then, we can explicitly write down the solutions for the phase differences, $\phi_j = \theta_{j+1} - \theta_j$:

$$H(\phi_j) = -g\frac{(N-j)j}{2}. \tag{15}$$

Since H is continuous and periodic, it is bounded; thus, there are phaselocked solutions only if g is sufficiently small. Since $H(x) = b$ generically has two roots, x, there are 2^N roots. We show that only one of these is asymptotically stable. Furthermore, we show that there is another fixed point with one positive eigenvalue and all other eigenvalues negative. The two branches of the one-dimensional unstable manifold form a closed invariant circle containing the stable fixed point. Figure 5B shows four fixed points on an invariant torus for $N = 2$. The filled fixed point is stable and the fixed point to the left of it has a one-dimensional unstable manifold forming the circle. Now, suppose that H has ± 1 as it extreme values. Then, from equation (15) $g < g^* \equiv N^2/8$ in order for phaselocking to occur. The righthand hand side is largest in magnitude when $j = N/2$. Thus, as g increases beyond g^*, there is a saddle-node bifurcation. We showed in [21] that the two fixed points on the aforementioned invariant circle merge and leave in their wake, an invariant circle. The frequency of this oscillations is roughly $q = a\sqrt{g - g^*}$ so that the phases, ϕ_j split into two groups. The frequency of the group with $j \leq N/2$ is higher than the group with $j > N/2$ by an amount q, thus the ideas provide a mechanism for frequency plateaus.

5.1.4 Two-Dimensions

The behavior of two-dimensional arrays, with coupling to only four the eight nearest neighbors turns out to be very similar to a cross product of behavior of two associated one-dimensional chains. Thus, in a sense, the equations are like separable PDEs. For simplicity, consider the network:

$$\theta'_{j,k} = \omega + H_N(\theta_{j,k+1} - \theta_{j,k}) + H_S(\theta_{j,k-1} - \theta_{j,k})$$
$$+ H_E(\theta_{j+1,k} - \theta_{j,k}) + H_W(\theta_{j-1,k} - \theta_{j,k}). \tag{16}$$

There are four different functions corrsponding to the four nearest neighbors. in [52], we analyzed this and generalizations for $N \times N$ arrays where N is large. We showed that

$$\theta_{j+1,k} - \theta_{j,k} \sim \Phi(j/N)$$
$$\theta_{j,k+1} - \theta_{j,k} \sim \Psi(j/N)$$

where $\Phi(x), \Psi(x)$ solved the corresponding horizontal and vertical chains. Thus, in a continuum approximation ($x = j/N, y = k/N$), $\theta_x = \Phi(x)$ and $\theta_y = \Psi(y)$ so that we can integrate this to obtain the full solution:

$$\theta(x, y) = \Omega t + \int_0^x \Phi(x') \, dx' + \int_0^y \Psi(y') \, dy'.$$

In particular, suppose that the medium is completely isotropic and $H_{N,S,E,W}(\phi) = H(\phi)$. Suppose that $H(\phi) = g(\phi) + C$ as in the end of section 5.1.2. Then we know that the phase gradient is linear so that we have:

$$\theta(x, y) = \Omega t + K[(x - \frac{1}{2})^2 + (y - \frac{1}{2})^2].$$

This representa a target wave; curves of constant phase are circles! In the other example from section 5.1.2, $H(\phi) = g(\phi + C)$, we have

$$\theta(x, y) = \Omega t + K[|x - \frac{1}{2}| + |y - \frac{1}{2}|],$$

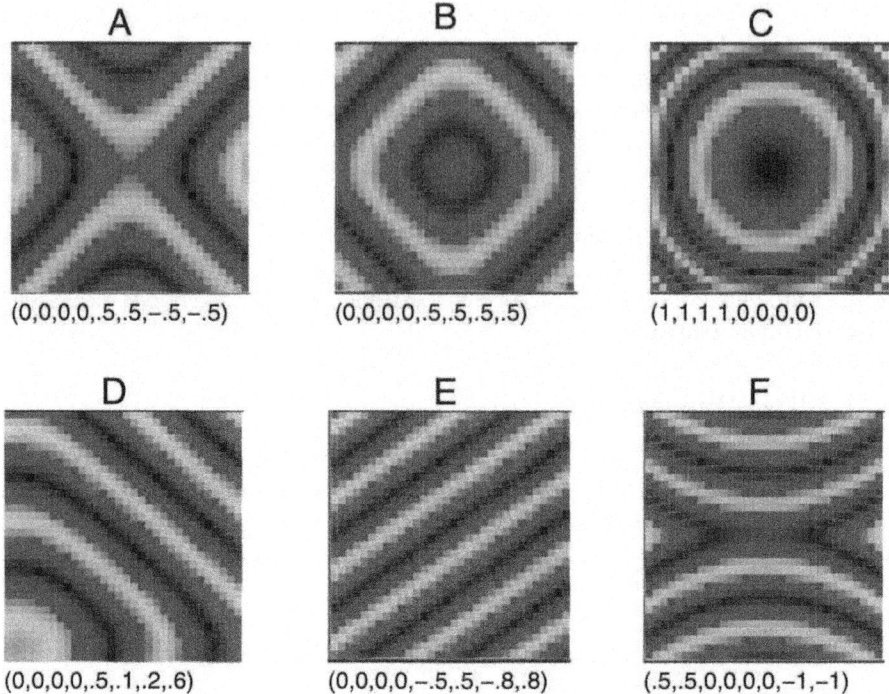

Fig. 6. Steady state phase relative to the upper-left oscillator in 32×32 arrays of locally coupled oscillators. Coupling is through a function of the form $A_0 + A_1 \cos\theta + B_1 \sin\theta$. Each of the four cardinal directions can be different. $B_1 = 1$ in all cases. Parameters in the parentheses correspond to $(A_0^E, A_0^W, A_0^S, A_0^N, A_1^E, A_1^W, A_1^S, A_1^N)$.

and waves are "square" target patterns. These and other patterns which are consequences of the theorem in [52] are illustrated in figure 6.

Spiral waves. The two-dimensional patterns described in the previous section are homotopic to the synchronous phaselocked solution and are a consequence of boundary inhomogeneities. Do there exist patterns of phases which are not branches of the synchronous state? For example, *in a ring*, the traveling wave solutions are topologically different from the synchronous solution and thus cannot be continued from that trivial branch. One analogue of such a pattern in two-dimensions is a spiral wave. Some of these waves are illustrated in figure 7. Figure 7A is a solution to the highly symmetric system:

$$\theta'_{jk} = \omega + \sum_{j',k'} \sin(\theta_{j',k'} - \theta_{jk})$$

where the sum is over the four nearest neighbors, {N,S,E,W}. In [49], we showed that there is a rotating wave-solution to this equation and with theorem 1, that it is asymptotically stable. Here is an example of the distribution of the phases for the 4×4 case:

$$
\begin{array}{cccc}
0 & \xi & \pi/2 - \xi & \pi/2 \\
-\xi & 0 & \pi/2 & \pi/2 + \xi \\
3\pi/2 + \xi & 3\pi/2 & \pi & \pi - \xi \\
3\pi/2 & 3\pi/2 - \xi & \pi + \xi & pi
\end{array}
$$

The reader is urged to check that there is a $\xi \in (0, \pi/4)$ which leads to a phaselocked solution and that this solution is stable. If we replace $H(u) = \sin u$ with the more general, $H(u) = \sin u + d(\cos u - 1)$, then the rotatinmg wave develops a twist and looks more like a spiral wave. Figure 7B shows such a spiral wave. As d increases, the spiral becomes tighter and tighter until eventually, the phase differences near the center become too great. The spiral disappears and leaves a "gently wobbling" spiral in its wake. Larger values of d result to large scale meandering of the spiral; a snapshot is shown in Figure 7C. Figure 7D shows another solution with the same parameters as in figure 7B, but with random initial data. There are many apparently stable steady state phaselocked solutions which consist of random arrangements of phase singularities.

6 Neural Networks

The previous sections dealt with networks of single neurons which were coupled together using chemical synapses. In this section, we are interested in firing-rate models; that is models of neurons in which the actual times of spikes are not specified. Rather, we specify the average firing rates of a neuron or a population of neurons. There are many different derivations of these firing rate models and we suggest that the reader consult [18, 15] for a general survey of formal means of obtaining the equations. We will present one method of deriving them that is tightly connected to the underlying biophysics of the neuron. Our method entails the use of a slow time scale in order to reduce a single neuron to a one (and possibly two) variable equation.

6.1 Slow Synapses

For simplicity, we consider a simple network with one excitatory cell which has slow spike-frequency adaptation (which we can always set to zero) with self-coupling and one inhibitory neuron which also is coupled to itself as well as to the excitatory cell. (The reason for self-coupling is to give the most general formulation of the reduction.) The derivation then suggests how to couple whole networks of neurons.

We start with the following system:

$$
C\frac{dV_e}{dt} = -I_{fast,e}(V_e, \ldots) - g_{ee}s_e(V_e - E_e) - g_{ie}s_i(V_e - E_i) \tag{17}
$$

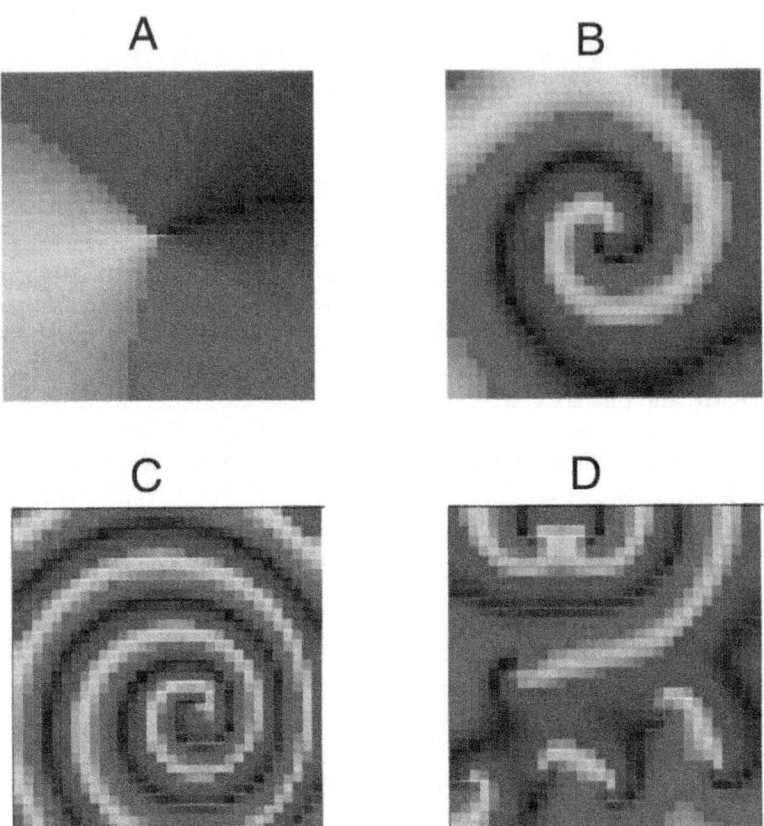

Fig. 7. Spiral waves in a locally coupled 32×32 oscillator array. (A) Pure sinusoidal coupling. (B) Coupling with $H(\theta) = \sin\theta + .6(1-\cos\theta)$. (C) With a cosine component of 0.8, locking is no longer possible and the core of the spiral wobbles as shown in this snapshot. (D) Same parameters as (B) but with random initial data.

$$- g_z z(V_e - E_z) + I_e$$

$$C\frac{dV_i}{dt} = -I_{fast,i}(V_i, \dots) - g_{ei}s_e(V_i - E_e) - g_{ii}s_i(V_i - E_i) + I_i \qquad (18)$$

$$\frac{ds_e}{dt} = \alpha_e(V_e)(1 - s_e) - s_e/\tau_e \qquad (19)$$

$$\frac{ds_i}{dt} = \alpha_i(V_i)(1 - s_i) - s_i/\tau_i \qquad (20)$$

$$\frac{dz_e}{dt} = \alpha_z(V_e)(1 - z_e) - z_e/\tau_z \qquad (21)$$

The first two equations represent the potentials of the excitatory and inhibitory cells respectively. The terms $I_{fast,*}$ may involve many additional variables such as the transient sodium and delayed rectifier channels necessary for spiking. The $s_{e,i}$ variables gate the synapses between the neurons and lie between 0 (fully closed) and 1 (fully open). The functions $\alpha_{e,i}$ are zero unless the neuron is spiking and then they are some finite value. $\tau_{e,i}$ are the time constants for the decay of the synapses. The parameters g_{jk} represent the maximal conductances of the synapses and $E_{e,i}$ are the reversal potentials of the excitatory and inhibitory synapses. Typically $E_e = 0$ mV and $E_i = -70$ mV although this latter value can be closer to -85 mV or up to -60 mV. The variable z gates the degree of spike-frequency adaptation for the excitatory cell; the function $\alpha_z(V)$ and the parameters, τ_z, g_z, and E_z are similar to those of the synapses.

In order to perform the needed reduction, we assume that $s_{e,i}$ and z are all much slower than the fast system. Thus, we can treat them as parameters. We assume that the fast dynamics of both excitatory and inhibitory cells is such that as current is increased, the neuron switches from a stable fixed point to large magnitude periodic firing. Furthermore, we also assume that the neuron is monostable so that there are no currents for which it can both stably oscillate and remain at rest. For example, if the fast-dynamics is class I, then as the current increases the fixed point diappears at a saddle-point and large-amplitude periodic oscillations emerge. Consider the excitatory dynamics:

$$\frac{dV_e}{dt} = -I_{fast,e}(V_e, \ldots) - G_{ee}(V_e - E_e) - G_{ie}(V_e - E_i) - G_z(V_e - E_z) + I_e$$

where G_{ee}, G_{ie}, G_z are parameters. Let

$$I(V; G_{ee}, G_{ie}, G_z, I_e) = -G_{ee}(V_e - E_e) - G_{ie}(V_e - E_i) - G_z(V_e - E_z) + I_e$$

be the applied current. Holding (G_{ee}, G_{ie}, G_z) constant, we suppose that as the applied current is increased, there is a saddle-node at $I_e^*(G_{ee}, G_{ie}, G_z)$. For $I_e < I_e^*$ the neuron is at rest and for $I_e > I_e^*$ the neuron fires repetitively with a frequency

$$f_e \approx \beta_e \sqrt{I_e - I_e^*}.$$

Now, we are ready for the reduction. Consider equation (19). If $I_e < I_e^*$ then the neuron is not firing and $\alpha_e(V_e) = 0$. If $I_e > I_e^*$, then the neuron fires repetitively with period, $T_e = 1/f_e$. We have assumed that this is fast compared to the dynamics of the synapse so that we can average the s_e dynamics obtaining

$$\frac{ds_e}{dt} = f_e \left[\int_0^{T_e} \alpha_e(V_e(t)) \, dt \right] (1 - s_e) - s_e/\tau_e.$$

(We have used the fact that $1/T_e = f_e$.) For many neural models, the spike-width is nearly independent of the frequency, so that the integral can be approximated as a constant independent of T_e, say, a_e. Using the frequency approximation, we see that

$$\frac{ds_e}{dt} = a_e \beta_e \sqrt{I_e - I_e^*}(1 - s_e) - s_e/\tau_e$$

when $I_e > I_e^*$. We recall that I_e^* is a function of (G_{ee}, G_{ie}, G_z) which are given by $(g_{ee}s_e, g_{ie}s_i, g_z z)$. Thus, to close the system, we need an expression for I_e^*. Numerically, for a wide variety of models, one finds that typically

$$I_e(G_{ee}, G_{ie}, G_z) \approx G_{ee}U_{ee} - G_{ie}U_{ie} - G_z U_z - I_{\theta,e}$$

where U_{ee}, U_{ie}, U_z are positive constants which have the dimension of potential and $I_{\theta,e}$ is the threshold current to initiate repetitive spiking in absence of any slow conductances. Assuming the fast dynamics of the excitatory and inhibitory cells are similar, we have reduced equations (17- 21) to the following three equations

$$\frac{ds_e}{dt} = c_e F(I_e + I_{ee}s_e - I_{ie}s_i - I_z z - I_{\theta,e})(1 - s_e) - s_e/\tau_e \qquad (22)$$

$$\frac{ds_i}{dt} = c_i F(I_i + I_{ei}s_e - I_{ii}s_i - I_{\theta,i})(1 - s_i) - s_i/\tau_i \qquad (23)$$

$$\frac{dz}{dt} = c_z F(I_e + I_{ee}s_e - I_{ie}s_i - I_z z - I_{\theta,e})(1 - z) - z/\tau_z, \qquad (24)$$

where $F(x) = \sqrt{\max(x,0)}$, $c_e = a_e\beta_e, c_z = a_z\beta_e, c_i = a_i\beta_i$ and $I_{ee} = g_{ee}U_{ee}$ and so on for the other I_{jk}.

Before continuing with our analysis of these models, we introduce a smoothed version of the function F. Numerous authors have shown that the presence of noise changes a sharp threshold for firing to a smoother and more graded firing rate. In a recent paper, [44] analyzed the firing properties of the normal form for a saddle-node. Using standard stochastic methods, they derive a complicated integral form for the firing rate curve as a function of the amount of noise. We find that a good approximation of such a curve is given by the smooth function:

$$F(I) = \sqrt{\log(1 + \exp(I/b))b}$$

where b characterizes the amount of "noise." That is, as $b \to 0^+$, F approaches the deterministic model. In the analysis below, we use $b = 0.1$.

6.2 Analysis of the Reduced Model

We sketch some possible types of solutions to the reduced network model. This is by no means exhaustive.

6.2.1 Purely Excitatory with No Adaptation

When there is no feedback inhibition and the adaptation is blocked, the equations reduce to

$$\frac{ds_e}{dt} = c_e F(I_e + I_{ee}s_e - I_{\theta,e})(1 - s_e) - s_e/\tau_e.$$

This is a one-dimensional system so that there are only fixed points. Since $F \geq 0$, the interval $[0, 1]$ is positively invariant so there exists at least one stable fixed point. Suppose, $I_{ee} = 0$ so there is no recurrent excitation. Then the fixed point is unique:

$$\bar{s}_e = \frac{c_e F(I_e - I_{\theta,e})}{c_e F(I_e - I_{\theta,e}) + 1/\tau_e}.$$

For small enough I_{ee} the fixed point persists and remains stable. Further increases in I_{ee} can lead to a fold bifurcation and the annihilation of the fixed point. But since we know there is always at least one fixed point, then there must in general (except at bifurcation points) be an odd number. Since this is a scalar system and the function F is continuous, the fixed points alternate in stability. For many choices of F, there is bistable behavior representing the neuron firing at a low or zero rate and at a much higher rate.

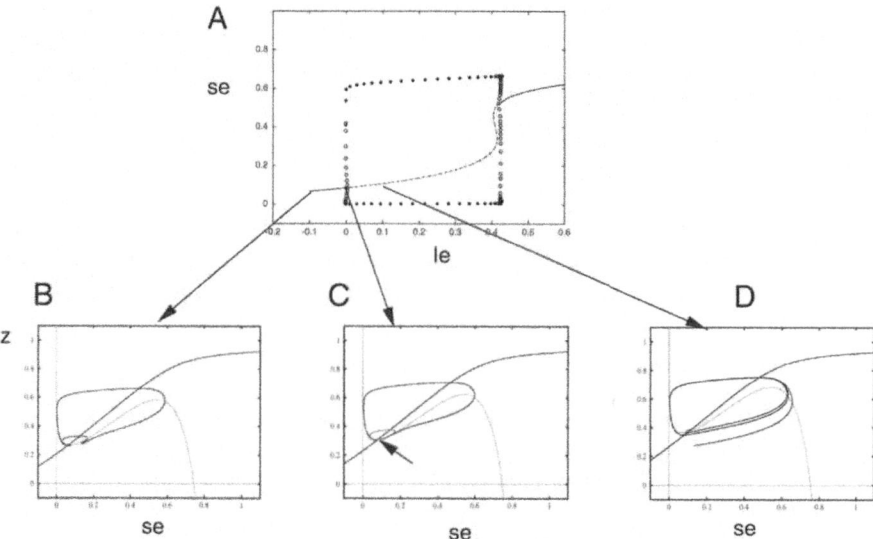

Fig. 8. Example behavior of a self-coupled excitary cell with adaptation. (A) The bifurcation diagram as the driving current, I_e increases. (B,C,D) three sample phase-planes at different driving currents showing excitability (B), bistability (C), and periodic orbits (D). In (C), arrow shows the stable fixed point and the small blue curve shows the unstable periodic orbit.

6.2.2 Excitatory with Adaptation

The presence of spike adaptation enables the network to oscillate and also allows it to be excitable. We point out that oscillations in s_e at the slow time scale assumed here correspond to bursting oscillations of the original model system. Baer et al [2] analyzed a slow two-variable system which modulates a fast spiking system through a saddle-node bifurcation and used this as a model for parabolic bursting. A rigorous analysis of a saddle-node that is slowly periodically driven was provided by Ermentrout and Kopell [22]. Figure 8 shows the bifurcation diagram for a set of parameters as the drive increases. Figure 8B shows the network in an excitable regime; weak stimuli lead to a decay to rest while stronger stimuli cause a burst of activity. Increasing the drive leads to a subcritical Hopf bifurcation and bistability between a large amplitude periodic orbit and a fixed point (Figure 8C). Further increases in the drive lead to a single stable periodic orbit (figure 8D).

6.2.3 Excitatory and Inhibitory

If we remove the adaptation, but allow inhibition, then we are in the case considered in the classic Wilson-Cowan model. This scenario has been explored thoroughly by many authors [3, 18]. The phaseplane possibilities are very close to those of the excitatory-adaptation model. There is more flexibility in the placement of the nullclines since the inhibitory cells receive inputs independent of the excitatory population. That is, there is the possibility for *feedforward* rather than feedback inhibition.

6.2.4 Full Model

Since the model is three variables, complex behavior is a possibility. For example if the adaptation is slow, then it can slowly move the excitatory-inhibitory "fast" system back and forth through bifurcations producing bursting [35].

6.3 Spatial Models

We can take the reduced model described in the previous section and create a spatially distributed network in order to understand the kinds of behavior in a one-dimensional slice of cortex. Consider the following spatial analogue of equations (22-24):

$$\frac{\partial s_e}{\partial t} = c_e F(I_e + I_{ee}J_{ee} * s_e - I_{ie}J_{ie} * s_i - I_z z - I_{\theta,e})(1 - s_e) - s_e/\tau_e \quad (25)$$

$$\frac{\partial s_i}{\partial t} = c_i F(I_i + I_{ei}J_{ei} * s_e - I_{ii}J_{ii} * s_i - I_{\theta,i})(1 - s_i) - s_i/\tau_i$$

$$\frac{\partial z}{\partial t} = c_z F(I_e + I_{ee}J_{ee} * s_e - I_{ie}J_{ie} * s_i - I_z z - I_{\theta,e})(1 - z) - z/\tau_z,$$

where

$$J * u \equiv \int_\Lambda J(x - y)u(y, t) \, dy.$$

In general, these spatial interaction functions are normalized so that the integral over the domain, Λ, is one. For the purposes of analysis, we will assume

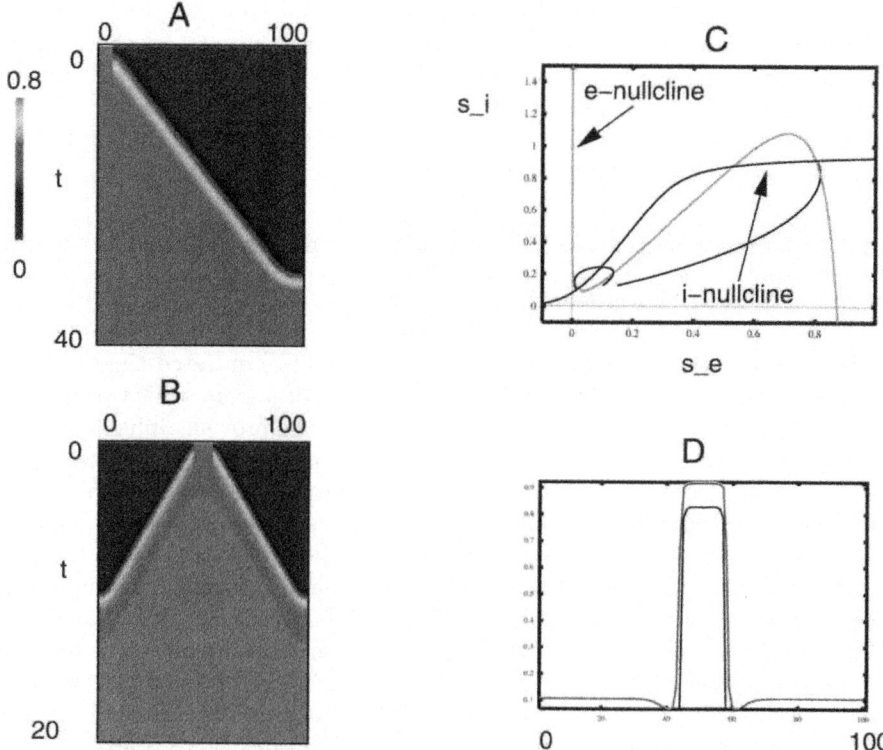

Fig. 9. Bistable media. (A) Space-time plot of a front for the scalar equation (26). Color scale is amplitude of $s_e(x, t)$ with time running vertically and space horizontally. $J_{ee} = \exp(-|x|/\sigma)/(2\sigma)$ with $\sigma = 2$. (B) Space-time plot for a network with excitation and inhibition with local dynamics as in (C). The spatial interactions are exponential with $\sigma_i = 1, \sigma_e = 2$. (C) Phaseplane for the $s_e - s_i$ system with no adaptation showing bistability. (D) Stationary localized bumps in the same parameter set as (B) but with excitation slowed by a factor of 2 and $\sigma_e = 1, \sigma_i = 2.5$. Horizontal axis indicates spatial index and the black and red curves are s_e, s_i respectively.

that the domain is the real line, a circle, the plane, or a torus. In the cases of a torus or a circle, the functions $J(x)$ will be periodic.

6.3.1 Bistability and Fronts

As in the previous section, we first consider the purely excitatory case so that the equations are simply:

$$\frac{\partial s_e}{\partial t} = c_e F(I_e + I_{ee} J_{ee} * s_e)(1 - s_e) - s_e/\tau_e. \qquad (26)$$

Suppose that $g(u) \equiv c_e F(I_e + I_e e u)(1 - u) - u/\tau_e$ has three roots. That is, suppose that the spatially homogeneous excitatory network is bistable. Then, as in the case of the bistable scalar reaction-diffusion equation, we expect that there may be front solutions joining one stable rest state to another. In fact, for the present model if we assume that J_{ee} is symmetric, positive, and has an integral of 1, then a theorem of Chen [8] implies that there exists a unique traveling front solution, $s_e(x, t) = S(x - ct)$ with velocity, c joining the two stable fixed points. We illustrate such a solution in figure 9A.

6.3.2 Fronts and Localized Excitation in the Presence of Inhibition

Figure 9B illustrates a front produced in a two-population network whose nullcline configuration is illustrated in figure 9C. Here we have turned off the adaptation so that $I_z = 0$. Unlike figure 9A, we have initiated the excitation in the center of the medium. Clearly, there is still a front which propagates but the rigorous existence of this front remains to be proven. Unlike the scalar system, however, the properties of bistability alone are not sufficient to guarantee a wave front. In particular, the time constant and the spatial extents matter in two-population models. Figure 9D illustrates a stationary localized solution to the excitatory-inhibitory network where we have slowed the excitation down by a factor of two and given the inhibitory cells a spatial extent two-and-a-half times that of the excitatory cells. Localized pulses have been the subject of numerous theoretical and numerical studies since the first rigororous analysis of this problem by Amari [1]. These pulses are thought to represent working or short-term memory [62]. Pinto and Ermentrout [50] used sigular perturbation (treating the extent of inhibition, σ_i, as a large parameter) to contruct solitary pulses in an excitatory-inhibitory network of the prsent form. Their starting point assumes the existence of fronts

6.3.3 Propagating Pulses

Suppose that we eliminate the inhibition. In experimental slice preparations, this is a common protocol in order to study the excitatory connectivity. We maintain the spike adaptation, z, and suppose that we have the phaseplane depicted in Figure 8A (and repeated in 10A). Then, instead of fronts as in figure 9A, we obtain traveling pulses as shown in Figure 10B. Essentially, if the adaptation is slow, then this can be regarded as a singular perturbation problem. For a fixed level of adaptation, the excitatory network is bistable and produces fronts between high and low levels of activity as in figure 9. Thus, this pulse can be viewed as a front and a "back" joined at a level of adaptation leading to the same speed. In other words, these localized traveling pulses are analogous to singular perturbation constructions of pulses in reaction-diffusion equations [7]. Indeed, Pinto and Ermentrout [51] use such a singular perturbation argument to construct these fronts.

In the scenario described above, the local dynamics is excitable; there is a single globally stable fixed point. However, the excitability of the medium depends on the slow dynamics of the adaptation. This type of excitability

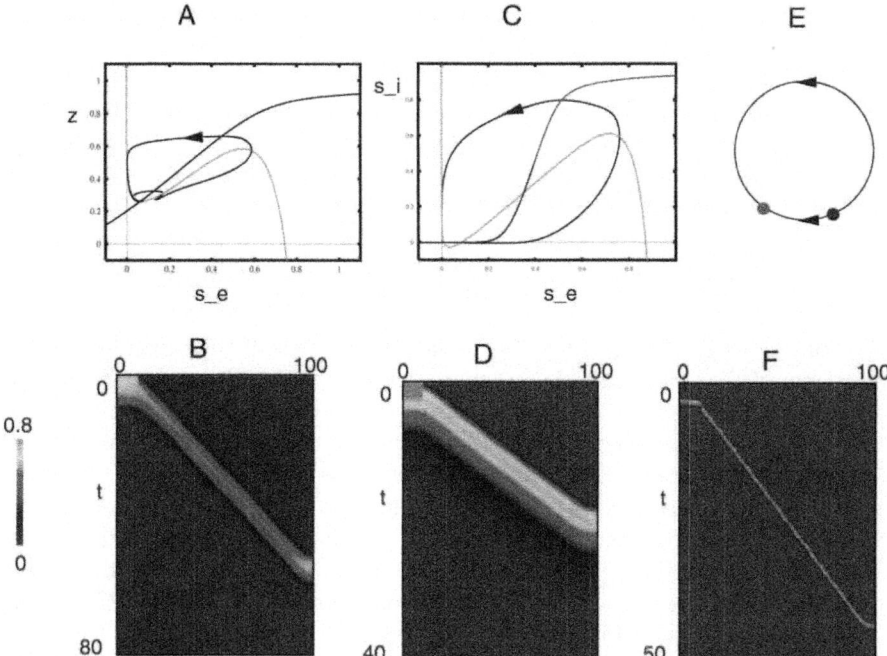

Fig. 10. Propagation of solitary pulses in an excitable network. (A) Phaseplane for a system with no inhibition showing excitable behavior. (B) Behavior of a network of 101 cells with exponentially decaying connectivity with local dynamics as in (A). $s_e(xt)$ is plotted with time increasing downward. (C) Excitable system with no adaptation and just inhibitory and excitatory cells. Note that there are three fixed points; only the leftmost one is stable. (D) Behavior of a network of excitatory and inhibitory cells where $\sigma_e = 4$ and $\sigma_i = 2$. (E) Ring model with two fixed point as a simple version for the dynamics in (C). (F) Solution to equation (27); $P(\theta)$ is plotted since this represents the actual activity.

is called Class II in the context of single neuron models ([54];see also figure 1B in the present article). Suppose that there is no adaptation, but instead, we allow inhibition. Then Class II excitability is still possible, but another type of excitability is now possible. Figure 10C shows the phaseplane for the excitatory-inhibitory network with no adaptation. Unlike the phaseplane in figure 10A, there are three fixed points. However, the rightmost fixed point is unstable. The middle fixed is a saddle-point whose unstable manifolds form a heteroclinic loop with the leftmost stable fixed point. Thus, we have the analogue of class I excitability as shown in figure 1C. The stable manifolds of the saddle point provides a true threshold; it is not necessary to have widely different timescales between the two variables to achieve excitability. Figure 10D shows the behavior of a network with excitatory and inhibitory neurons whose local dynamics is that of 10C. There has been little analysis of this type of network excitability. However, we offer a simple approximation leading to a scalar problem which may be amenable to analysis. The dynamics of the pulse is largely confined to a tubular neighborhood of the unstable manifolds of the saddle-point which form a ring. This suggests a scalar "ring model" for the local dynamics as shown in figure 10E. Ermentrout and Rinzel [27] used such a ring model to study propagation in a reaction-diffusion excitable system. Consider the following system:

$$\frac{\partial \theta(x,t)}{\partial t} = F[\theta(x,t)] + gR[\theta(x,t)] \int_\Lambda J(x-y)P[\theta(y,t)] \, dy. \qquad (27)$$

The functions F, R, P are, say, 2π−periodic functions of their arguments. The function: $q(u) = F(u) + R(u)P(u)$ characterizes the ring dynamics. We assume that there are two fixed points, u_s and u_t representing the stable fixed point and the threshold respectively. Since q is periodic, there is also a fixed point, $u_s + 2\pi$. The function R represents the response of the ring model to inputs; since this is a class I system, R will usually be non-negative. The function $P(u)$ is the shape of the pulse of activity; it is positive and narrowly peaked. We assume that $R(u_s) > 0$ so that inputs to a resting cell will excite it past threshold. We can ask whether the scalar problem, equation (27), admits a traveling *front* solution joining u_s to $u_s + 2\pi$. Note that since the phase space is a ring, this actually represents a traveling pulse. Osan et al [48] studied a similar system but in the context of single neurons and with a time-dependent synapse replacing the function $P(u)$. Figure 10E shows a simulation of a network of 101 cells with $J(x)$ exponential, $R(u) = 1 + \cos u$, $P(u) = (.5 + .5 \cos(u - 2.5))^5$, $F(u) = .98 - 1.02 \cos(u)$ and $g = 2.2$. Since θ is a front, we plot $P(\theta)$ instead since these are the real coordinates. Despite the similarity of (27) to (26), we cannot apply Chen's theorem to the former. The reason for this is that one of the assumptions of the Chen theorem is equivalent to $P'(u) \geq 0$. Since P is a pulse-like function, this assumption will be violated. The precise conditions for the existence of solutions like figure 10E for (27) remain to be determined.

6.3.4 Oscillatory Networks

Consider the excitatory-inhibitory network with no adaptation. Suppose that we are in the same local configuration as figure 10C and drive the excitatory cells enough so that the lower two fixed points merge and disappear. Then, the local network oscillates. Figure 11A shows the phaseplane of the EI system with no adaptation. Now suppose that we couple the network with exponentially decaying functions for which the excitation spreads more than the inhibition ($\sigma_e > \sigma_i$). Then, as seen in figure 11B, the network generates waves which eventually synchronize. However, if the inhibition spreads farther than the excitation, the synchronous solution is no longer the unique stable solution. One such non-synchronous solution is illustrated in figure 11C; the

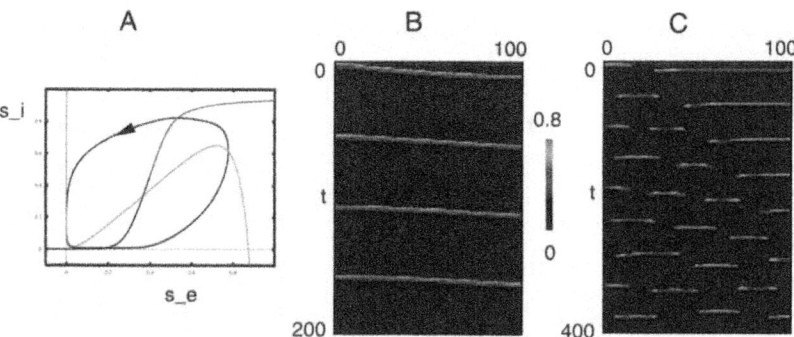

Fig. 11. Oscillatory E-I network. (A) Phaseplane for the local dynamics showing the existence of a stable limit cycle. (B) A network where $\sigma_e = 2$ and $\sigma_i = 1$. s_e, s_i are initialized at identical low value. The first 10 excitatory cells are raised to a value of 1 inducing a wave of activity. The network oscillates and tends to a synchronous solution. (C) Same as (B), but σ_i is raised to 4 leading to an instability of the synchronous state and the appearance of clustered solutions

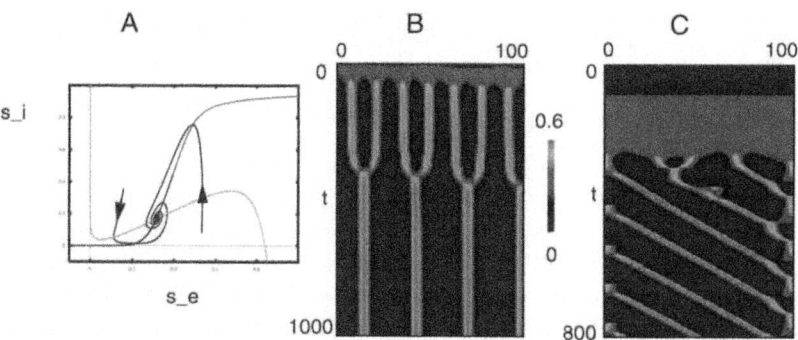

Fig. 12. The Turing instability. (A) The phaseplane for the local dynamics; there is a stable fixed point on the middle branch. (B) The same system generates spatial stationary patterns due to the long-range inhibition; $\sigma_e = 2.0, \sigma_i = 7.5$. (C) Some of the inhibition is removed and an equal amount of adaptation is added to compensate. The rest state of the local dynamics is still stable. However, the same spatial interactions as in (B) lead to a destabilization of the rest state through a Hopf bifurcation and a nonzero wave number. The spatial patterns become traveling waves.

network breaks up into clustered states. Other initial data lead to synchrony so that this network seems to be bistable.

6.3.5 Turing-Type Instabilities

As a final example of spatial effects on networks, we consider a system where *all* of the interesting dynamics arises due to the spatial interactions. That is, the local dynamics consists of a single stable fixed point with no excitability. We now connect this up with long-range inhibition and short-range excitation. This is the classical scenario for pattern formation and results in an instability of the uniform state to a nonzero wave number. We sketch the basic ideas of the analysis. Let $\bar{s}_e, \bar{s}_i$ be a spatially uniform fixed point to equation (25) when there is no adaptation. Write $s_e = \bar{s}_e + u$ and $s_i = \bar{s}_i + v$. The linearized system has the form

$$\frac{\partial}{\partial t}\begin{pmatrix} u \\ v \end{pmatrix} = \begin{pmatrix} -c_{ee}u + d_{ee}J_{ee} * u & -d_{ie}J_{ie} * v \\ d_{ei}J_{ei} * u & -c_{ii}v - d_{ii}J_{ii} * v \end{pmatrix}$$

Since the rest state is on the middle branch, this means that $-c_{ee} + d_{ee} > 0$ so that $d_{ee} > 0$. All of the other coefficients are obviously non-negative. Suppose that we work on the infinite domain. Then solutions to this linear problem have the form:

$$[u(x,t), v(x,t)] = e^{ikx + \lambda(k)t}[\bar{u}, \bar{v}]$$

where $k \geq 0$. $\{\lambda(k); [\bar{u}, \bar{v}]\}$ is a eigenvalue-eigenvector pair for the matrix:

$$M(k) \equiv \begin{pmatrix} -c_{ee} + d_{ee}\hat{J}_{ee}(k) & -d_{ie}\hat{J}_{ie}(k) \\ d_{ei}\hat{J}_{ee}(k) & -c_{ii} - d_{ii}\hat{J}_{ii}(k) \end{pmatrix}$$

where

$$\hat{J}(k) = \int_{-\infty}^{\infty} J(x)e^{-ikx} \, dx.$$

We note that if J is an exponential or Gaussian, then $\hat{J}$ is non-negative and decreases with $|k|$. If $M(k)$ has eigenvalues with positive real parts for some k, then the rest state is unstable and perturbations with this wavelength will tend to grow. This sets the template for patterns for the full nonlinear system [46]. Since M is a 2×2 matrix, stability is guaranteed if the trace is negative and the detrminant positive. If we suppose that $J_{ee} = J_{ei} = J_e$ and $J_{ie} = J_{ii} = J_i$, then, the trace, δ and the determinant, Δ satisfy:

$$\delta(k) = -(c_{ii} + c_{ee}) + d_{ee}\hat{J}_e(k) - d_{ii}\hat{J}_i(k)$$
$$\Delta(k) = c_{ee}c_{ii} + (d_{ie}d_{ei} - d_{ii}d_{ee})\hat{J}_e(k)\hat{J}_i(k) - c_{ii}d_{ee}\hat{J}_e(k) + c_{ee}d_{ii}\hat{J}_i(k).$$

At $k = 0$, we know that the rest state is stable and for k large, all $k-$dependent terms vanish so that the rest state is stable to large k perturbations. However, because $d_{ee} > 0$ it is possible for intermediate values of k to produce instabilities. This point is critical: if the rest state is not on the increasing middle-branch of the E-nullcline, then no spatial instabilities are possible. If the self-inhibition is weak, then d_{ii} is small and the trace will be negative for $k > 0$ if it is negative for $k = 0$ as assumed. Consider for simplicity the case where $d_{ii} = 0$. Then the trace is a monotonically decreasing function of k and since it is assumed to be negative when $k = 0$, the trace is always negative. On the other hand, the determinant can become negative if d_{ee} is large enough since the term $\hat{J}_e(k)\hat{J}_i(k)$ decreases more rapidly than $\hat{J}_e(k)$. Such a scenario can then lead to an instability as an eigenvalue crosses through zero at some non-zero k. Figure 12B illustrates as example of this instability for the full nonlinear system. Two-dimensional analogues of this instability were suggested as a mechanism for visual hallucinations in Ermentrout and Cowan [19].

Suppose that we include adaptation as well as inhibition. We can remove some of the inhibition to compensate for the presence of adaptation which acts like a local slow inhibition. Figure 12C shows the result of this; instead of stationary patterns of activity, we obtain traveling waves. What is surprising about these waves is that the domain is finite and the boundaries seem to have little influence on the waveform.

References

1. Amari, S, 1977, Dynamics of patternformation in lateral-inhibition type neural fields, Biol. Cybernet. 27:77-87.
2. Baer, S.M., J. Rinzel, and H. Carrillo: Analysis of an autonomous phase model for neuronal parabolic bursting. J. Math. Biol. 33:309-333 (1995).
3. Beer, R. On the Dynamics of Small Continuous-Time Recurrent Neural Networks, Adaptive Behavior, 3:469-509, 1994.
4. P.D. Brodfuehrer, E.A. Debski, B.A. O'Gara, and W.O. Friesen, 1995, Neuronal control of leech swimming, J. Neurobiology, 27:403-418.
5. Buck, J., 1988, Synchronous rhythmic flashing in fireflies, II. Q. Rev. Biol. 63:265-289.
6. P.L. Buono and M. Golubitsky. Models of central pattern generators for quadruped locomotion: I. primary gaits. J. Math. Biol. 42 (4) (2001) 291-326.
7. Casten RG, Cohen H, Lagerstrom PA (1975) Perturbation analysis of an approximation to the Hodgkin-Huxley theory. *Quart. J. Appl. Math.* 32(4):365-402.
8. Chen, X, Existence, Uniqueness, and Asymtotic Stability of Traveling Waves in Nonlocal Evolution Equations, Advances in Differential Equations 2, 125-160 (1997)
9. Chen, Z and Ermentrout, B Wave Propogation Mediated by $GABA_B$ Synapse and Rebound Excitation in an Inhibitory Network: A Reduced Model Approach, Journal of Computational Neuroscience 5, 53-69 (1998).
10. C.C. Chow and N. Kopell, 'Dynamics of spiking neurons with electrical coupling', Neural Comp. **12**, 1643-1678 (2000).
11. Chow, CC. Phaselocking in weakly heterogeneous neuronal networks. Physica D 118:343-370 (1998).
12. A.H. Cohen, S. Rossignol, and S. Grillner, 1988, Neural Control of Rhythmic Movements in Vertebrates, New York, Wiley.
13. Cohen AH, Ermentrout GB, Kiemel T, Kopell N, Sigvardt KA, Williams TL. Modelling of intersegmental coordination in the lamprey central pattern generator for locomotion. Trends Neurosci. 1992 Nov; 15(11):434-8.
14. Crook, S and A Cohen. (1995) Central pattern generators. In Bower and Beeman, editors, The Book of GENESIS: A workbook of tutorials for the GEneral NEural SImulation System, Chapter 6.

15. Dayan, P. and Abbott, L. Theoretical Neuroscience, 2001, MIT Press, Cambridge MA, Chapt 7.
16. Ermentrout,-G.-Bard, Stable periodic solutions to discrete and continuum arrays of weakly coupled nonlinear oscillators. SIAM-J.-Appl.-Math. 52 (1992), no. 6, 1665-1687.
17. Ermentrout-B, Linearization of F-I curves by adaptation. Neural-Comput. 1998 Oct 1; 10(7): 1721-9
18. Ermentrout-B, Neural networks as spatio-temporal pattern-forming systems, Reports on Progress in Physics, 61:353-430, 1998.
19. Ermentrout,-G.-B.; Cowan,-J.-D., A mathematical theory of visual hallucination patterns. Biol.-Cybernet. 34 (1979), no. 3, 137-150.
20. Ermentrout-B; Flores-J; Gelperin-A, Minimal model of oscillations and waves in the Limax olfactory lobe with tests of the model's predictive power. J-Neurophysiol. 1998 May; 79(5): 2677-89
21. Ermentrout, G.B. Kopell, N. Frequency plateaus in a chain of weakly coupled oscillators. I. 1984 SIAM-J.-Math.-Anal. 15 (1984), no. 2, 215–237.
22. Ermentrout,-G.-B.; Kopell,-N., Parabolic bursting in an excitable system coupled with a slow oscillation. SIAM-J.-Appl.-Math. 46 (1986), no. 2, 233-253.
23. Ermentrout G.B. and Kopell, N 1991, Multiple pulse interactions and averaging in systems of coupled neural oscillators, J. Math. Biology 29:195-217
24. Ermentrout GB, Kopell N. Fine structure of neural spiking and synchronization in the presence of conduction delays. Proc Natl Acad Sci U S A. 1998 Feb 3;95(3):1259-64.
25. Ermentrout GB, Kleinfeld D Traveling electrical waves in cortex: insights from phase dynamics and speculation on a computational role NEURON 29: (1) 33-44 JAN 2001
26. Ermentrout B, Pascal M, Gutkin B The effects of spike frequency adaptation and negative feedback on the synchronization of neural oscillators NEURAL COMPUT 13 (6): 1285-1310 JUN 2001
27. Ermentrout,-G.-Bard; Rinzel,-John, Waves in a simple, excitable or oscillatory, reaction-diffusion model. J.-Math.-Biol. 11 (1981) 269-294.
28. Ermentrout B, Wang JW, Flores J, and Gelperin, A. Model for olfactory discrimination and learning in Limax procerebrum incorporating oscillatory dynamics and wave propagation J NEUROPHYSIOL 85 (4): 1444-1452 APR 2001
29. W. Gerstner, J.L. van Hemmen, and J. Cowan. What matters in neuronal locking? Neural Comp. **8** 1653-1676 (1996).
30. Goel P, Ermentrout B Synchrony, stability, and firing patterns in pulse-coupled oscillators PHYSICA D 163 (3-4): 191-216 MAR 15 2002
31. C.M. Gray, 1994, Synchronous oscillations in neuronal systems: mechanisms and functions, J. Computat. Neurosci. 1:11-38.
32. D. Hansel, G. Mato, and C. Meunier. Synchrony in excitatory neural networks. Neural Comp. **7**, 307-337 (1995).
33. Hanson FE. Comparative studies of firefly pacemakers. Fed Proc. 1978 Jun;37(8):2158-64.
34. Hodgkin, AL (1948) The local changes associated with repetitive action in a non-medulated axon. J. Physiol (London) 107: 165-181
35. Izhikevich E.M. (2000) Neural Excitability, Spiking, and Bursting. International Journal of Bifurcation and Chaos. 10:1171-1266.
36. Izhikevich E.M. (2003) Simple Model of Spiking Neurons IEEE Transactions on Neural Networks, in press

37. Kopell,-N.; Ermentrout,-G.-B., Symmetry and phaselocking in chains of weakly coupled oscillators. Comm.-Pure-Appl.-Math. 39 (1986), no. 5, 623–660.
38. Kopell,-N.; Ermentrout,-G.-B., Phase transitions and other phenomena in chains of coupled oscillators. 1990 SIAM-J.-Appl.-Math. 50 (1990), no. 4, 1014–1052.
39. N. Kopell and G.B. Ermentrout. "Mechanisms of phaselocking and frequency control in pairs of coupled neural oscillators", for Handbook on Dynamical Systems, vol. 2: Toward applications. Ed. B. Fiedler, Elsevier, pp 3-5. (2002)
40. Kopell N, Ermentrout GB, Whittington MA, Traub RD. Gamma rhythms and beta rhythms have different synchronization properties. Proc Natl Acad Sci U S A. 2000 Feb 15;97(4):1867-72.
41. Y. Kuznetsov, Elements of Applied Bifurcation Theory, Springer, NY, 2000, page 83.
42. Lam YW, Cohen LB, Wachowiak M, Zochowski MR. Odors elicit three different oscillations in the turtle olfactory bulb. J Neurosci. 2000 Jan 15;20(2):749-62.
43. P.E. Latham, B.J. Richmond, P.G. Nelson, and S. Nirenberg. Intrinsic dynamics in neuronal networks. I. Theory. J. Neurophysiol. 83(2):808-827 (2000).
44. Lindner, B. and Longtin, A. Neural Computation, (in press) 2003
45. E. Marder and R.L. Calabrese, 1996, Principles of rhythmic motor pattern generation, Physiological Reviews, 687-717.
46. Murray, J.D. Mathematical Biology, Springer NY 1989
47. Nenadic Z, Ghosh BK, Ulinski Propagating waves in visual cortex: a large-scale model of turtle visual cortex. J Comput Neurosci. 2003 Mar-Apr;14(2):161-84.
48. Osan R, Rubin J, Ermentrout B Regular traveling waves in a one-dimensional network of theta neurons SIAM J APPL MATH 62 (4): 1197-1221 APR 22 2002
49. Paullet, J.E and Ermentrout, G.B. Stable rotating waves in two-dimensional discrete active media. SIAM-J.-Appl.-Math. 54 (1994), no. 6, 1720–1744.
50. Pinto, D.J. and Ermentrout, G.B. 2001a, Spatially Structured Activity in Synaptically Coupled Neuronal Networks: II. Lateral Inhibition and Standing Pulses, SIAM J. Appl Math. 62(1):226–243
51. Pinto, D.J. and Ermentrout, G.B. 2001b, Spatially Structured Activity in Synaptically Coupled Neuronal Networks: I. Traveling Fronts and Pulses SIAM J. Appl Math. 62(1):206–225
52. Ren, LW and Ermentrout, G.B. Monotonicity of phaselocked solutions in chains and arrays of nearest-neighbor coupled oscillators. SIAM-J.-Math.-Anal. 29 (1998), no. 1, 208–234
53. Reyes AD, Fetz EE, (1993a) Effects of transient depolarizing potentials on the firing rate of cat neocortical neurons. J Neurophysiol 1993 May;69(5):1673-83
54. Rinzel-J; Ermentrout-B, Analysis of neural excitability and oscillations, In "Methods in Neuronal Modelling: From synapses to Networks", C. Koch and I. Segev, eds. 1989, MIT Press (revised 1998).
55. Jonathan Rubin and David Terman, "Geometric Singular Perturbation Analysis of Neuronal Dynamics," in B. Fiedler, editor, Handbook of Dynamical Systems, vol. 3: Toward Applications, Elsevier, 2002.
56. Schoner G, Jiang WY, Kelso JA. A synergetic theory of quadrupedal gaits and gait transitions. J Theor Biol. 1990 Feb 9;142(3):359-91.
57. W. Singer, 1993, Synchronization of cortical activity and its putative role in information processing and learning, Ann Rev. Physiol, 55:349-374.
58. M. Stopfer, S. Bhagavan, B.H. Smith, and G. Laurent, 1997, Impaired odour discrimination on desynchronization of odour-encoding neural assemblies, Nature 390:70-74.

59. Traub RD, Jefferys JG, Whittington MA. Simulation of gamma rhythms in networks of interneurons and pyramidal cells. J Comput Neurosci. 1997 Apr;4(2):141-50.
60. Traub RD, Whittington MA, Stanford IM, Jefferys JG. A mechanism for generation of long-range synchronous fast oscillations in the cortex. Nature. 1996 Oct 17;383(6601):621-4.
61. Van Vreeswijk C, Abbott L, and Ermentrout B. When inhibition not excitation synchronizes neural firing. J Comput Neurosci. 1994 Dec; 1(4):313-21.
62. Wang X.-J. (1999a) Synaptic basis of cortical persistent activity: the importance of NMDA receptors to working memory. J. Neurosci. 19, 9587-9603.
63. White JA, Chow CC, Ritt J, Soto-Trevino C, and Kopell N. Synchronization and oscillatory dynamics in heterogeneous, mutually inhibited neurons. J. Comp. Neurosci. 5, 5-16 (1998).
64. Winfree, A.T. Biological rhythms and the behavior of populations of coupled oscillators, J. Theoret. Biol. 16:15-42, 1967.

Physiology and Mathematical Modeling of the Auditory System

Alla Borisyuk

Mathematical Biosciences Institute, Ohio State University
W. 18th Avenue 231, 43210-1292 Ohio, USA
borisyuk@mbi.osu.edu

1 Introduction

It is hard to imagine what the world would be like if it was put on "mute". For most people the vast and diverse stream of auditory information is an indispensable part of the environment perception. Our auditory system, among other things, allows us to understand speech, appreciate music, and locate sound sources in space. To be able to perform all perceptual functions the auditory system employs complex multi-stage processing of auditory information. For example, the sound waves are separated into frequency bands, which later in the system are often fused to create perception of pitch, small amplitude and frequency modulations are detected and analyzed, timing and amplitude differences between the ears are computed, and everything is combined with information from other sensory systems.

In this chapter we will discuss how the nervous system processes auditory information. We will need to combine some knowledge from anatomy (to know which structures participate), electrophysiology (that tells us the properties of these structures), and, finally, mathematical modeling (as a powerful tool for studying the mechanisms of the observed phenomena). In fact, one of the goals of this chapter is to show examples of types of mathematical modeling that have been used in the auditory research.

The material in this chapter is based on lectures given by Catherine Carr and Michael Reed at the Mathematical Biosciences Institute, Ohio State University in April 2003. It also uses material from the following books: "The Mammalian auditory pathway: neurophysiology" (A.N. Popper, R.R. Fay, eds. [77]), "Fundamentals of hearing: an introduction" by W.A. Yost [116], "An introduction to the psychology of hearing" by B.C.J. Moore [67], "From sound to synapse" by C. Daniel Geisler [26], chapter "Cochlear Nucleus" by E.D. Young and D. Oertel in [90]; and websites:
<http://serous.med.buffalo.edu/hearing/>,
<http://www.neurophys.wisc.edu/aud/training.html>.

1.1 Auditory System at a Glance

Let us start by looking at the overall path that the auditory information travels through the brain. The auditory system differs significantly from the visual and somatosensory pathways in that there is no large direct pathway from peripheral receptors to the cortex. Rather, there is significant reorganization and processing of information at the intermediate stages. We will describe these stages in more details in the subsequent sections, but for now just outline the overall structure.

Sound waves traveling through the air reach the ears — the outer part of the peripheral auditory system. Then they are transmitted, mechanically, through the middle ear to the auditory part of the inner ear — the cochlea. In the cochlea the mechanical signal is converted into the electrical one by auditory receptors, namely, hair cells. This transduction process is fundamental in all sensory systems as external signals of light, chemical compounds or sounds must be "translated" into the language of the central nervous system, the language of electrical signals.

Next, the signal originating from each cochlea is carried by the auditory nerve into the brainstem (see Figs. 1, 2). Figure 3 shows a 3-dimensional reconstruction of the human brainstem. It consists of three main parts: pons, medulla, and midbrain. The auditory nerve first synapses in the cochlear nuclei inside the medulla. From the cochlear nuclei, auditory information is split into at least two streams. One stream projects directly to the auditory part of the midbrain, inferior colliculus (Fig. 3); while the other goes through another set of nuclei in the medulla, called superior olivary complex.

The first of these pathways is thought to pick up the tiny differences in acoustic signals, that you need, for example, to differentiate between similar words. The indirect stream, which is the one that we are mostly going to consider, originates in the ventral cochlear nucleus. Auditory nerve fibers that bring information to this stream end with giant hand-like synapses. This tight connection allows the timing of the signal to be preserved up to a microsecond (which is very surprising, because the width of action potential, the "unit signal", is on the order of milliseconds). This precisely timed information is carried on to the superior olive, both on the same side of the brain and across midline. In the superior olive the small differences in the timing and loudness of the sound in each ear are compared, and from this you can determine the direction the sound is coming from. The superior olive then projects up to the inferior colliculus.

From the inferior colliculus, both streams of information proceed to the sensory thalamus (Fig. 1-2). The auditory nucleus of the thalamus is the medial geniculate nucleus. The medial geniculate projects to the auditory cortex, located in the temporal lobes inside one of the folds of the cortex (Fig. 4).

It is important to notice that many cells at each level of auditory system are *tonotopically organized*. This means that in each of the auditory structures

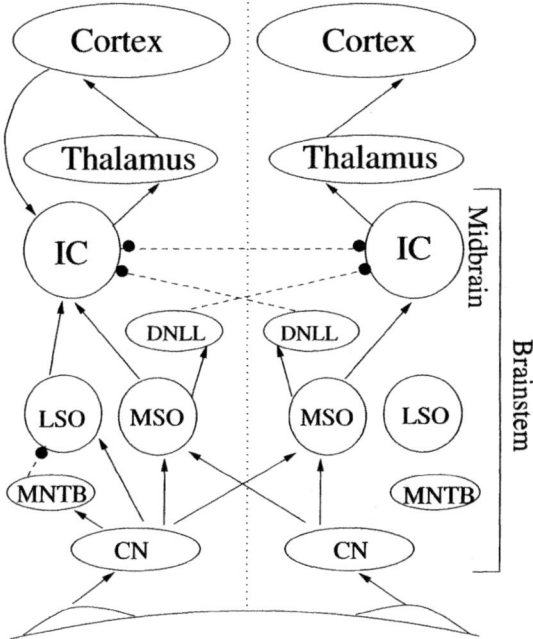

Fig. 1. Schematic of some of the auditory pathways from the ears (at the bottom) to the cortex. Most of the auditory nuclei appear on both sides of the brain and are shown twice: to the right and to the left of the (dotted) midline. Abbreviations are as follows: CN – cochlear nucleus; MNTB – medial nucleus of the trapezoidal body; MSO – medial superior olive; LSO – lateral superior olive; DNLL – dorsal nucleus of the lateral lemniscus; IC – inferior colliculus. Excitatory connections are shown with solid lines; inhibitory connections — with dashed.

it is possible to identify areas in which cells preferentially respond to sounds of certain frequencies (tones) and the preferred frequencies vary more-or-less continuously with position of the cell.

Notice also that both pathways starting at the cochlear nuclei are bilateral. The consequence of this is that localized brain lesions anywhere along the auditory pathway usually have no obvious effect on hearing. Deafness is usually caused by damage to the middle ear, cochlea, or auditory nerve. In infants the auditory system continues to develop throughout the first year of life. Neurons in the brainstem mature and many connections are just beginning to form (e.g. between brainstem nuclei, thalamic nuclei, and auditory cortex). As for other sensory systems, stimulation during this time is essential for normal development. When hearing is impaired in early life, the morphology and function of auditory neurons may be affected.

The overall structure of the sub-cortical auditory system and even some anatomical details are similar across many species. A lot of information about the auditory system has been obtained over the years from experiments, es-

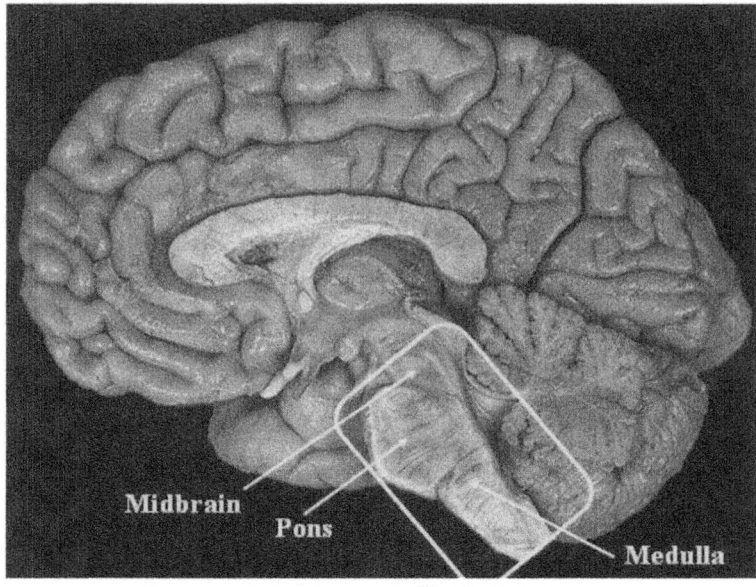

Brainstem

Fig. 2. Location of brainstem in the human brain. Three main parts of the brainstem are indicated with arrows. (Picture of the brain is from Digital Anatomist Project [20], Department of Biological Structure, University of Washington, with permission).

pecially anatomical and electrophysiological, with cats, bats, gerbils, owls, monkeys and many other animals. Also, a large number of behavioral (psychophysical) studies have been conducted with animals as well as humans.

1.2 Sound Characteristics

Here we describe some of the physical characteristics of the auditory stimuli and some of the perceptual characteristics that we ascribe to them.

Sounds are pressure waves that transfer energy and momentum from the source to places around it. One of their natural characteristics is the amount of energy transmitted by a wave in unit time, called the *power of the wave*. The power of a wave is proportional to the square of the frequency, the square of the amplitude and the wave speed. Often it is convenient to consider the power per unit area of the wavefront, i.e. the amount of energy transmitted by the wave perpendicularly to its wavefront in unit time per unit area. This is called the *intensity of a wave*. The sound waves can also be characterized by specifying their time-dependent spectra. The way these physical characteristics of sounds are usually described and quantified relates to our perception of the sounds.

The spectral properties of many sounds evoke a sensation of *pitch*. Pitch is defined as the auditory attribute on the basis of which tones may be ordered on

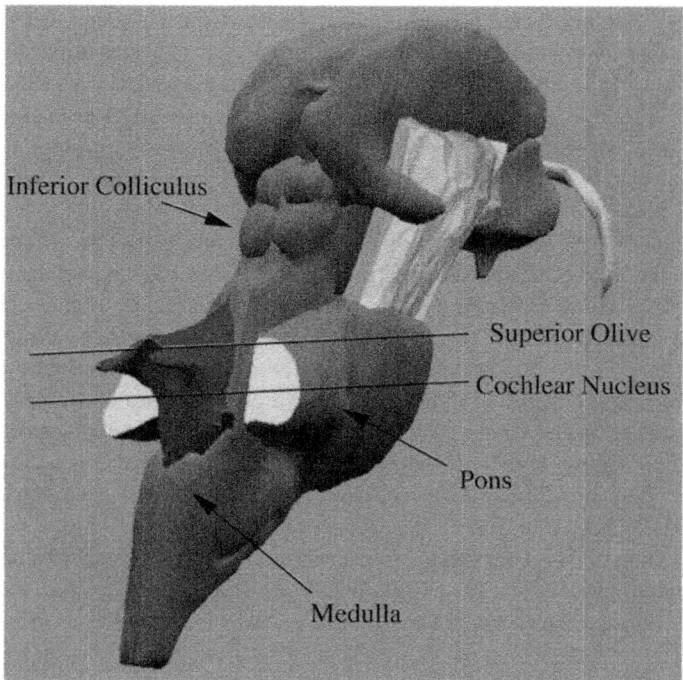

Fig. 3. Reconstructed view of human brainstem (red), together with thalamus (purple) and hypothalamus (yellow) (From Digital Anatomist Project [20], Department of Biological Structure, University of Washington, with permission). Arrows point different parts of the brainstem, while lines indicate planes within which cochlear nuclei and superior olives are located.

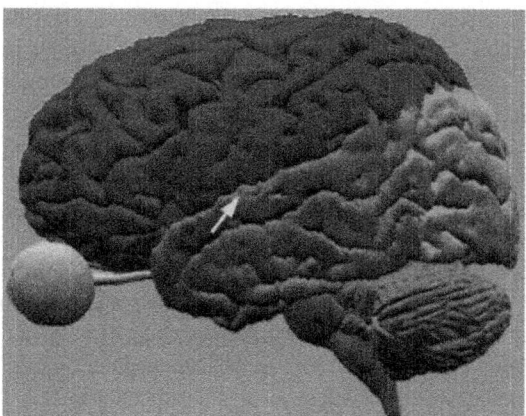

Fig. 4. Lobes of the brain (From Digital Anatomist Project [20], Department of Biological Structure, University of Washington, with permission): frontal lobe (blue), parietal lobe (green), temporal lobe (purple), occipital lobe (yellow). White arrow points at the fold within which most of the auditory cortex is located.

a musical scale. The pitch of a tone is related to its frequency or periodicity. The pitch of a periodic sound wave (simple tone) is usually indicated by specifying its frequency in Hz. The pitch of a complex tone is usually indicated by the frequency of a simple tone whose pitch would be perceived to be the same. A simple tone evokes a sensation of pitch only if its frequency is between 20 Hz and 5 kHz, while the spectrum of audible frequencies extends from 20 Hz to 20 kHz.

As a side note: in musical and psychophysical literature there are two different aspects of the notion of pitch. One is related to the frequency of a sound and is called *pitch height*; the other is related to the place in a musical scale and is called *pitch chroma*. Pitch height corresponds to the sensation of 'high' and 'low'. Pitch chroma, on the other hand, describes the perceptual phenomenon of octave equivalence, by which two sounds separated by an octave (and thus relatively distant in terms of pitch height) are nonetheless perceived as being somehow equivalent. Thus pitch chroma is organized in a circular fashion. Chroma perception is limited to the frequency range of 50-4000 Hz [122].

The intensity of the sound is commonly quantified on the logarithmic scale in Bels, or decibels. In this logarithmic scale the intensity I of a sound is quantified relative to the intensity I_0 of a reference sound, and is measured in Bels:

$$1 \text{ Bel} = \log_{10}\left(\frac{I}{I_0}\right),$$

or in decibels (dB):

$$1 \text{ dB} = \frac{1}{10}\log_{10}\left(\frac{I}{I_0}\right).$$

These expressions give the difference between the intensities in Bels or dB. For example, if I is hundred times I_0, then the level of I is 2 Bel or 20 dB greater than that of I_0. If we want to express an absolute intensity I using decibels then we have to use some standard reference intensity I_0. The reference intensity most commonly used is 10^{-12} (Watt/m^2), which corresponds to a pressure of 20 μPa or about $2 \cdot 10^{-10}$ atmosphere. The sound level of a sound relative to this standard intensity is called the *sound pressure level* or SPL of the sound. This standard reference intensity was chosen because it is very close to the faintest sound of 1 kHz that a human can detect. This lower limit to the detectable sounds is set by the Brownian motion of the molecules in the ear. They produce noise input to the auditory receptors, so the signal has to be strong enough to be detected on top of the noise. Notice that the sound pressure level can be negative, and that 0 dB sound does not mean the absence of a sound wave.

The reason for using the logarithmic scale is that when the sound intensity increases linearly from low to high, it creates the perception that the "loudness" (i.e. the perceived intensity) increases fast at first and then the increase slows down as the sound gets louder. Logarithmic scale is also useful because of the great range of intensities that the auditory system deals with. Dynamic range of human hearing is approximately 140 dB, ranging from a whisper to a jet engine. To give some examples, the sound level at an average home is about 40 dB, an average conversation is about 60 dB, and a loud rock band is about 110 dB.

2 Peripheral Auditory System

Peripheral auditory system serves to transfer sounds from the environment to the brain. It is composed of three main components – outer, middle, and inner ear (Fig. 5).

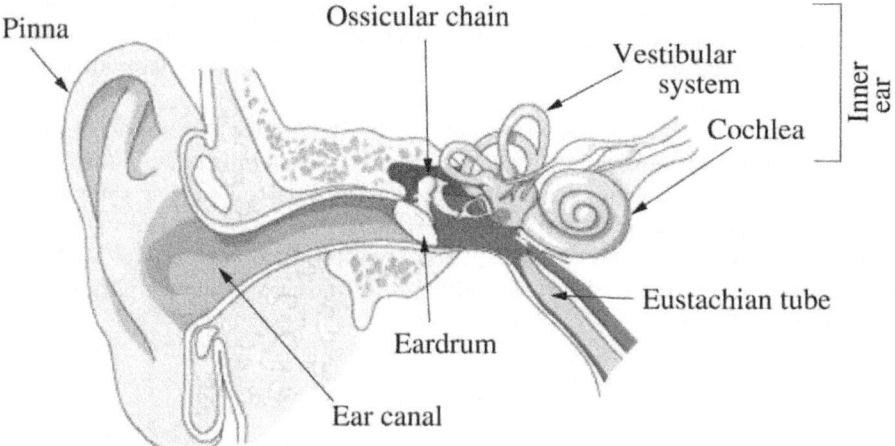

Fig. 5. Peripheral auditory system. From <http://www.workplacegroup.net/article-ear-anatomy.htm>. With permission from W.C. Earnshaw at The Wellcome Trust Institute foe Cell Biology, Edinburgh.

2.1 Outer Ear

The outer ear is the part that we see — *the pinna* (a structure made of cartilage that we call "the ear" in common speech) and *the ear canal*. A sound wave travelling in the air, before it reaches the entrance of the ear canal, must interact with the head, the torso, and the intrinsic shapes of the pinna. As a result, the waves can be amplified or attenuated, or their spectrum may be modified. These changes depend on the specific frequencies of the waves (e.g., they are most significant for waves of frequency above 1.5 kHz), but also on the three-dimensional position of the sound source. Our brain can use the difference of spectra between the two ears to determine the position of the original sound source. A typical way to characterize the pressure that an arbitrary sound produces at the eardrum is by using the, so called, Head-Related

Transfer Functions (HRTFs). These are constructed by placing microphones at the entrance of the subject's ear canal, and recording the generated pressure as sounds originate from point sources at various locations. The resulting functions, or their Fourier transforms – HRTF, are surprisingly complicated functions of four variables: three space coordinates and frequency. Much research has been devoted to measurements, and computations of the external ear transformations, both for human and animal subjects (see, for example, [66],[89],[109] for humans, [68] for cats), and databases can be found on the web, for example at <http://www.ircam.fr/equipes/salles/listen/>, or <http://interface.cipic.ucdavis.edu/CIL_html/CIL_HRTF_database. htm>. The HRTFs are used in many engineering applications, most popularly in creation of realistic acoustic environments in virtual reality settings: games, virtual music halls, etc. Unfortunately, because of the diversity of the pinnae shapes and head sizes, HRTFs vary significantly from one person to another. For this reason, each of the existing models of acoustic fields, based on some average HRTF, is found adequate by only a relatively small number of listeners.

Having been collected and transformed by the pinnae, the sound waves are directed into the ear canal, where they travel towards *the eardrum* (also called *the tympanic membrane*). The ear canal maintains the proper temperature and humidity for the elasticity of the eardrum. It also contains tiny hairs that filter dust particles, and special glands that produce earwax for additional protection. The ear canal can also resonate sound waves and amplify tones in the 3000-4000 Hz range.

2.2 Middle Ear

The eardrum (tympanic membrane), and the neighboring cavity with three tiny bones (*ossicular chain*) comprise the middle ear (Fig. 5). The sound waves travelling through the ear canal cause the vibrations of the tympanic membrane, and these vibrations are further transmitted through the chain of bones towards the inner ear. For the eardrum to function optimally, it should not be bent either inwards or outwards in the absence of the sound waves. This means that the air pressure should be the same in the outer and middle ear. The pressure equalization is achieved through the *eustachian tube* — a tube which connects the middle ear cavity with the back of the throat. Normally this tube is closed, but it opens with swallowing or shewing, thus working as a pressure equalizer. The middle ear bones (malleus, incus and stapes) are the smallest bones in the body. They work together as a lever system, to amplify the force of the vibrations. The malleus is attached to the tympanic membrane, the stapes enters the oval window of the inner ear, and the incus lies in between.

2.3 Inner Ear. Cochlea. Hair Cells.

The inner ear is a series of fluid-filled structures. It has two main parts: a part that is involved in balance and is called *vestibular system*, and a second part that is responsible for converting sounds from mechanical vibrations into electrical signals, and is called *cochlea*. Cochlea, in mammals, is coiled as a snail's shell (Fig. 5). It has two membrane-covered openings into the middle ear: the *oval window* and the *round window*, and a membranous sac (containing receptor cells) that separates them (Fig. 6). The size of the oval window is 15 to 30 times smaller than that of the eardrum. This size difference produces amplification needed to match impedances between sound waves in the air and in the cochlear fluid. The principal function of the membranous cochlear sac is to act as a hydromechanical frequency analyzer.

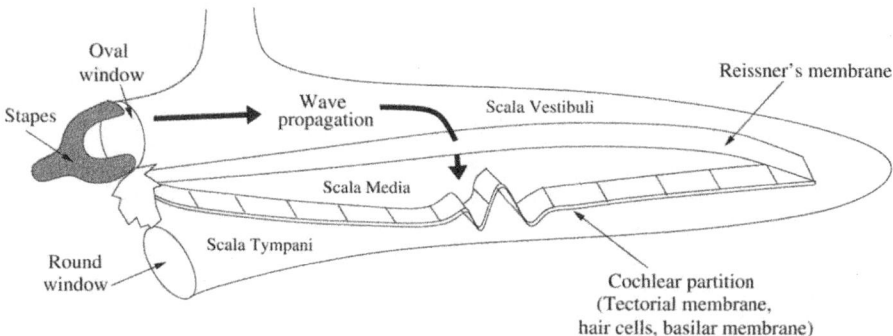

Fig. 6. Schematic of a mammalian cochlea, straightened (see text).

Here is a rough outline of how it works (see Figure 6). The input from the middle ear arrives at the oval window and creates pressure fluctuations in the fluid. These fluctuations travel along the cochlea and eventually are dissipated by the movements of the large round window, which serves as a pressure release for incompressible fluids. As the pressure waves travel along the membraneous sac, they permeate one of the walls (*Reissner's membrane*) and cause vibrations in the other (*cochlear partition*). The platform of the cochlear partition (*basilar membrane*) changes in its mechanical properties along its length from being narrow and stiff at the base of the cochlea, to being wide and compliant at the apex. Therefore, the lower the frequency of the tone the further from the oval window the vibration pattern is located.

Hair Cells

Looking at higher resolution inside the cochlear partition (Fig. 7A), there are auditory sensory cells (*inner and outer hair cells*) that sit on the basilar membrane and have their stereocilia sticking out into the fluid and attached to a floppy *tectorial membrane*. Because the basilar and tectorial membranes attach to the bone at different points, their motion causes the tectorial membrane to slide across the basilar membrane, tilting the hair bundle (Fig. 7B). Using electrophysiological recordings and mechanical stimulation, it has been demonstrated that deflection of the stereocilia leads to a change of the membrane potential of the cell. Deflection towards the highest stereocilia leads to a depolarization of the cell. It is also shown that both the initial change in charge and calcium concentration occur near the tips of the stereocilia, suggesting that the transduction channels are located at the tips. Moreover, the calcium entry and charge change happens relatively fast (10 μsec), which suggests that the transduction channels are gated mechanically rather than through a second messenger system.

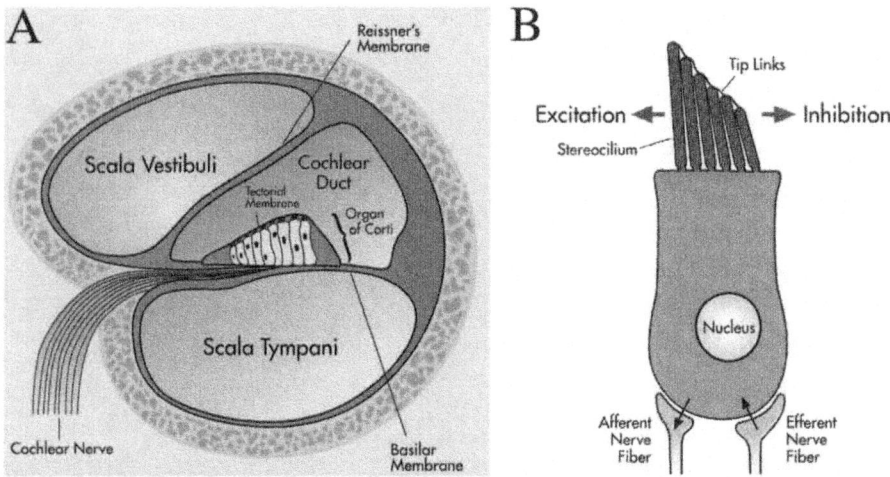

Fig. 7. Location and basic structure of hair cells. *A:* Radial segment of cochlea showing main components of the cochlear partition. *B:* Schematic of a hair cell. Bold arrows indicate that excitation and inhibition of the cell voltage can be produced by deflections of the hair bundle (see text). Illustration is made by Brook L. Johnson.

There are two types of auditory sensory cells: one row of inner hair cells (IHCs; they are called "inner" because they are closer to the bony central core of the twisted cochlea), and three to five rows of outer hair cells (OHCs). Inner and outer cells have different functions, and different sets of incoming and outgoing connections. Each inner hair cell's output is read out by twenty or so *cochlear afferent neurons of type I* (for characteristics of types of afferent neurons see below), and each type I afferent neuron only contacts a single inner hair cell. The output of each outer hair cell, on the other hand, is read together with many other outer hair cell outputs, by several *type II afferent neurons*.

All afferent neurons send their axons to the cochlear nuclei of the brainstem and terminate there. Outer hair cells also receive descending (efferent) signals from neurons with cell bodies in the brainstem. The afferent and efferent axons together form the auditory nerve — the communication channel between the peripheral auditory system and the auditory regions of the brain.

Inner hair cells convey frequency, intensity and phase of signal. As explained above, to a first approximation, the frequency is encoded by the identity of the activated inner hair cells, i.e. those inner hair cells that are located at the appropriate place along the cochlea. The intensity of the signal is encoded by the DC component of the receptor potential; and the timing — by the AC component (see *Phase locking* below).

The function of the outer hair cells is not presently known. Two main theories are that they act as cochlear amplifiers, and that they act to affect the movement of the tectorial membrane. Another theory is that they function as motors, which alter the local micromechanics to amplify the sensitivity and frequency selectivity of the cochlea. It seems that the outer hair cells participate in the mechanical response of the basilar membrane, because loss of outer hair cells leads to a loss of sensitivity to soft sounds and decrease in the sharpness of tuning.

2.4 Mathematical Modeling of the Peripheral Auditory System

Mathematical and computational modeling of the peripheral auditory system has a long history. For mathematical analysis and models of the external and middle ear see, e.g. review by Rosowski [84]. More recently, there has been a number of three-dimensional models of the middle ear that use the finite-element method (e.g. [44, 27, 18, 22, 25, 41, 103]). Some of these models have been successfully used for clinical applications (e.g. [41, 18]).

An even larger set of studies concentrated on the modeling of the cochlea (for reviews see, for example, [40, 28]. Early models represented cochlea as one- or two-dimensional structure and incorporated only a few elements of the presently known cochlear mechanics. Early one-dimensional models of the cochlea [23, 76] have assumed that the fluid pressure is constant over a cross-section of the cochlear channel. The fluid is assumed to be incompressible and inviscid, and the basilar membrane is modeled as a damped, forced harmonic oscillator with no elastic coupling along its length. Qualitatively, this model has been shown to capture the basic features of the basilar membrane response. Quantitatively, however, it yields large discrepancies with measurement results [120]. Two-dimensional models by Ranke [78] and Zwislocki [121] make similar assumptions on the cochlear fluid and the basilar membrane. Ranke's model uses a deep water approximation, while Zwislocki used the shallow water theory in his model. These models were further developed in [3, 4, 53, 92] and in other works. Other two-dimensional models incorporate more sophisticated representations of the basilar membrane using, for example, elastic beam and plate theory [10, 17, 36, 37, 43, 99]. Three-dimensional models were considered by Steele and Taber [100] and de Boer [19], who used asymptotic methods and computations and obtained an improved fit of the experimental data. Their work seems to indicate that geometry may play a significant role in the problem. In particular, the effect of the spiral coiling of the cochlea on the wave dynamics remains unresolved; see [101, 105, 56, 60].

With the development of more powerful computers it became possible to construct more detailed computational models of the cochlea. A two-dimensional computational model of the cochlea was constructed by Beyer [7]. In this model the cochlea is a flat rectangular strip divided into two equal halves by a line which represents the basilar membrane. The fluid is modelled by the full Navier-Stokes equations with a viscosity term, but elastic coupling along the basilar membrane is not incorporated. Beyer has used a modification of Peskin's immersed boundary method, originally developed for modeling the fluid dynamics of the heart [75]. Several three-dimensional computational models have been reported, such as Kolston's model [45], intended to simulate the micro-mechanics of the cochlear partition in the linear regime (i.e., near the threshold of hearing), Parthasarati, Grosh and Nuttal's hybrid analytical-computational model using WKB approximations and finite-element methods, and Givelberg and Bunn's model [28] using the immersed boundary method in a three-dimensional setting.

We will consider as an example the analysis of a two-dimensional model with fluid viscosity, by Peskin [73, 74]. In this model cochlea is represented by a plane and the basilar membrane by an infinite line dividing the plane into two halves. The fluid in this model satisfies the Navier-Stokes equations with the non-linearities dropped. The distinctive feature of this model is that the location of wave peak for a particular sound frequency is strongly influenced by fluid viscosity and the negative friction of the basilar membrane. This model was studied with asymptotic and numerical methods in [54, 55]. In the example that we present here, the zeroth order approximation of the solution is found by WKB method. (The WKB method was first used for this cochlear model by Neu and Keller [69] in the case of zero membrane friction).

Problem setup: Basilar membrane at rest is situated along the x-axis. The deviation of the basilar membrane from the axis will be denoted by $h(x, t)$. Vector (u, v) is the velocity of the fluid, and p is the pressure of the fluid (these are functions of (x, y, t)).

Assumption 1: the fluid satisfies the Navier-Stokes equations with the non-linearities left out, i.e.

$$\text{for } y \neq 0 :$$

$$\rho \frac{\partial u}{\partial t} + \frac{\partial p}{\partial x} = \mu \left(\frac{\partial^2 u}{\partial x^2} + \frac{\partial^2 u}{\partial y^2} \right), \tag{1}$$

$$\rho \frac{\partial v}{\partial t} + \frac{\partial p}{\partial y} = \mu \left(\frac{\partial^2 v}{\partial x^2} + \frac{\partial^2 v}{\partial y^2} \right), \tag{2}$$

$$\frac{\partial u}{\partial x} + \frac{\partial v}{\partial y} = 0. \tag{3}$$

In these equations, ρ is the density and μ is the viscosity of the fluid. The last equation represents the incompressibility of the fluid.

Notice that there is no explicit forcing term in this model. Yet, there are bounded wave solutions that move in the direction of the increasing x as if there is a source of vibration at $x = -\infty$.

Assumption 2:
a) the basilar membrane (located at $y = 0$) has zero mass;
b) there is no coupling along the membrane;
c) each point of the membrane feels a restoring force that is proportional to the displacement and to the compliance (flexibility) of the membrane, the latter given by $e^{\lambda x}$ (to represent the fact that the actual membrane is more narrow and more flexible at the far end);
d) the membrane possesses an active mechanism (mechanical amplifier, parameter β below is negative), i.e.

for $y = 0$:

$$u(x, 0, t) = 0, \tag{4}$$

$$v(x, 0, t) = \frac{\partial h}{dt}(x, t), \tag{5}$$

$$p(x, 0^-, t) - p(x, 0^+, t) = s_0 e^{-\lambda x} \left(h + \beta \frac{\partial h}{\partial t} \right)(x, t). \tag{6}$$

Note that the boundary conditions are applied at the rest position of the membrane $y = 0$, not at its instantaneous position $h(x, t)$. This, as well as the above assumptions is justified by the small displacements of the membrane and the fluid particles in the cochlea. Notice also that there is an unknown function $h(x, t)$ in the boundary conditions. Finding this function such that the rest of the system has a bounded solution (u, v, p) is part of the problem. In addition, the problem is complicated by presence of the stiffness term. The parameter λ was measured (for dead cochlea) by von Békésey: $\lambda^{-1} \approx 0.7$ cm, i.e. over the length of the human cochlea (length ≈ 3.5 cm) the stiffness more than doubles.

Because of the symmetry of the system, we look for solutions that satisfy

$$p(x, y, t) = -p(x, -y, t),$$

$$u(x, y, t) = -u(x, -y, t),$$

$$v(x, y, t) = v(x, -y, t).$$

Then we can restrict the system of equations (1-3) to $y < 0$ and, using notation $y = 0$ instead of $y = 0^-$, re-write the boundary conditions (4-6) as

$$u(x, 0, t) = 0, \tag{7}$$

$$v(x, 0, t) = \frac{\partial h}{dt}(x, t), \tag{8}$$

$$2p(x, 0, t) = s_0 e^{-\lambda x} \left(h + \beta \frac{\partial h}{\partial t} \right)(x, t). \tag{9}$$

We also impose the condition

$$(u, v, p) \to 0 \text{ as } y \to -\infty.$$

Solution plan: We want to determine the solution that represents the steady state response of the cochlea to a pure tone. We will look for this solution as a small perturbation from the time-periodic function of the same frequency as the tone (and thus introduce a small parameter in the system). Further, we will use asymptotic expansion in the small parameter and by solving the zeroth order system of equations we will find an approximation of the solution.

We use a change of variables

$$\begin{pmatrix} u \\ v \\ p \end{pmatrix}(x, y, t, \epsilon) = \begin{pmatrix} U \\ V \\ P \end{pmatrix}(x - x_\varepsilon, y/\varepsilon, \varepsilon) e^{i\left(\omega t + \frac{\Phi(x - x_\varepsilon)}{\varepsilon}\right)},$$

$$h(x, t, \varepsilon) = H(x - x_\varepsilon, \varepsilon) e^{i\left(\omega t + \frac{\Phi(x - x_\varepsilon)}{\varepsilon}\right)},$$

where Φ is the local spatial frequency, ω is the given frequency of the external pure tone stimulus (radians/second), functions U, V, P, H, Φ are complex-valued, and the parameters ε and x_ε will be chosen later on. We set

$$X = x - x_\varepsilon,$$

$$Y = y/\varepsilon,$$

and

$$\xi(X) = \Phi'(X) = \frac{\partial \Phi}{\partial X}(X).$$

In terms of the new variables:

$$\frac{\partial u}{\partial t} = i\omega \cdot U \cdot e^{i\left(\omega t + \frac{\Phi(x - x_\varepsilon)}{\varepsilon}\right)},$$

$$\frac{\partial u}{\partial x} = \left[\frac{\partial U}{\partial X} + \frac{i}{\varepsilon} \xi(X) U \right] e^{i\left(\omega t + \frac{\Phi(x - x_\varepsilon)}{\varepsilon}\right)},$$

$$\frac{\partial u}{\partial y} = \frac{1}{\varepsilon} \frac{\partial U}{\partial y} e^{i\left(\omega t + \frac{\Phi(x - x_\varepsilon)}{\varepsilon}\right)},$$

$$\Delta u = \Delta U =$$

$$= \left[\frac{\partial^2 U}{\partial X^2} + \frac{i}{\varepsilon} \left(\xi' U + 2\xi \frac{\partial U}{\partial X} \right) + \frac{1}{\varepsilon^2} \left(\frac{\partial^2 U}{\partial Y^2} - \xi^2 U \right) \right] e^{i\left(\omega t + \frac{\Phi(x - x_\varepsilon)}{\varepsilon}\right)}.$$

Then the equations (1-3) can be rewritten as:

for $Y < 0$:

$$(i\omega\rho - \mu\Delta)U + \left(i\frac{\xi}{\varepsilon} + \frac{\partial}{\partial x}\right)P = 0, \tag{10}$$

$$(i\omega\rho - \mu\Delta)V + \frac{1}{\varepsilon}\frac{\partial P}{\partial Y} = 0, \tag{11}$$

$$\left(i\frac{\xi}{\varepsilon} + \frac{\partial}{\partial x}\right)U + \frac{1}{\varepsilon}\frac{\partial V}{\partial Y} = 0, \tag{12}$$

with boundary conditions

for $Y = 0$:

$$
\begin{aligned}
U(X,0,\varepsilon) &= 0, \\
V(X,0,\varepsilon) &= i\omega H(X,0,\varepsilon), \\
2P(X,0,\varepsilon) &= s_0(1 + i\omega\beta)e^{-\lambda x_\varepsilon}e^{-\lambda x}H(X,0,\varepsilon), \\
(U,V,P) &\to 0 \quad \text{as} \quad Y \to -\infty.
\end{aligned}
$$

Assumption 3. The functions U, V, P, H have the expansion

$$U = U_0 + \varepsilon U_1 + \cdots,$$

$$V = V_0 + \varepsilon V_1 + \cdots,$$

$$P = \varepsilon(P_0 + \varepsilon P_1 + \cdots),$$

$$H = H_0 + \varepsilon H_1 + \cdots.$$

We now choose

$$\varepsilon^2 = \frac{\mu\lambda^2}{\rho\omega}, \quad e^{-\lambda x_\varepsilon} = \varepsilon.$$

This choice of ε makes it indeed a small parameter for realistic values of other quantities. For example, if the fluid has characteristics of water ($\rho = 1$ g/cm^3, $\mu = .02$ g/(cm· s)), for a 600 Hz tone ($\omega = 2\pi\cdot600$/s) and $1/\lambda=.7$ cm, $\varepsilon \approx .003$.

If we substitute the expansions of U, V, P, H into the equations (10-12) and collect terms with matching powers of ε, we find at the zeroth order:

for $Y < 0$:

$$\rho\omega\left(i - \frac{1}{\lambda^2}\left(-\xi^2 + \frac{\partial^2}{\partial Y^2}\right)\right)U_0 + i\xi P_0 = 0,$$

$$\rho\omega\left(i - \frac{1}{\lambda^2}\left(-\xi^2 + \frac{\partial^2}{\partial Y^2}\right)\right)V_0 + \frac{\partial P_0}{\partial Y} = 0,$$

$$i\xi U_0 + \frac{\partial V_0}{\partial Y} = 0,$$

and for $Y = 0$:

$$U_0(X, 0) = 0,$$

$$V_0(X, 0) = i\omega H_0(X),$$

$$2P_0(X, 0) = s_0(1 + i\omega\beta)e^{-\lambda x} H_0(X),$$

$$(U_0, V_0, P_0) \to 0 \text{ as } Y \to -\infty.$$

Next, at the first order

for $Y < 0$:

$$\rho\omega \left(i - \frac{1}{\lambda^2} \left(-\xi^2 + \frac{\partial^2}{\partial Y^2} \right) \right) U_1 + i\xi P_1 = \frac{\rho\omega}{\lambda^2} i \left(\xi' + 2\xi \frac{\partial}{\partial X} \right) U_0 - \frac{\partial}{\partial X} P_0,$$

$$\rho\omega \left(i - \frac{1}{\lambda^2} \left(-\xi^2 + \frac{\partial^2}{\partial Y^2} \right) \right) V_1 + \frac{\partial P_1}{\partial Y} = \frac{\rho\omega}{\lambda^2} i \left(\xi' + 2\xi \frac{\partial}{\partial X} \right) V_0,$$

$$i\xi U_1 + \frac{\partial V_1}{\partial Y} = -\frac{\partial}{\partial X} U_0,$$

and for $Y = 0$:

$$U_1(X, 0) = 0,$$

$$V_1(X, 0) = i\omega H_1(X),$$

$$2P_1(X, 0) = s_0(1 + i\omega\beta)e^{-\lambda X} H_1(X),$$

$$(U_1, V_1, P_1) \to 0 \text{ as } Y \to -\infty.$$

Consider the zeroth order equations. For each fixed x, the functions P_0, U_0, V_0 satisfy an ODE system whose solution is given by

$$P_0(X, Y) = P_0(X, 0)e^{\sqrt{\xi^2} Y},$$

$$U_0(X, Y) = -P_0(X, 0)\frac{i\xi}{i\omega\rho} \left(e^{\sqrt{\xi^2} Y} - e^{\sqrt{\xi^2 + i\lambda^2} Y} \right),$$

$$V_0(X, Y) = -P_0(X, 0)\frac{\xi^2}{i\omega\rho} \left(\frac{e^{\sqrt{\xi^2} Y}}{\sqrt{\xi^2}} - \frac{e^{\sqrt{\xi^2 + i\lambda^2} Y}}{\sqrt{\xi^2 + i\lambda^2}} \right),$$

where $\sqrt{}$ denotes the root with positive real part.

Next, the zeroth order equations that contain H_0 give two different formulae for H_0 in terms of $P_0(X, 0)$:

$$H_0(X) = \frac{V_0(X, 0)}{i\omega} = P_0(X, 0)\frac{\xi^2}{\omega^2\rho} \left(\frac{1}{\sqrt{xi^2}} - \frac{1}{\sqrt{\xi^2 + i\lambda^2}} \right) \tag{13}$$

and

$$H_0(X) = P_0(X,0)\frac{2e^{\lambda X}}{s_0(1 + i\omega\beta)}. \tag{14}$$

We have just expressed our zeroth order solution $U_0(X,Y)$, $V_0(X,Y)$, $P_0(X,Y)$, $H_0(X)$ in terms of $P_0(X,0)$ and $\xi(X)$. Now there are two steps remaining. Step 1 is to find a function $\xi(X)$ such that both equations for $H_0(X)$ are satisfied. Step 2 is to find $P_0(X,0)$.

Step 1. For ξ to satisfy both (13) and (14) we need to have

$$Q(X) = \xi^2\left(\frac{1}{\sqrt{\xi^2}} - \frac{1}{\sqrt{\xi^2 + i\lambda^2}}\right), \tag{15}$$

where

$$Q(X) = \frac{2\rho\omega^2 e^{\lambda X}}{s_0(1 + i\omega\beta)}.$$

The condition (15) is called a dispersion relation. It implicitly defines the local spatial frequency ξ in terms of X and ω. Equation (15) may have multiple solutions, for example, if ξ is a solution, then $-\xi$ is also a solution (a wave going in the opposite direction). Using notation $\eta = \sqrt{\xi^2}$ we can transform (15) into

$$\eta^3 - \frac{1}{2}(Q + \frac{i\lambda^2}{Q})\eta^2 + i\lambda^2\eta - \frac{1}{2}i\lambda^2 Q = 0.$$

But not every zero of this cubic satisfies the dispersion relation. First of all the real part of η has to be non-negative. Second, by definition of Q: $Q < \eta$, therefore only the zeros of the cubic such that

$$Q = \eta - \frac{\eta^2}{\sqrt{\eta^2 + i\lambda^2}}$$

satisfy the original equation. For every root of cubic that satisfies these two conditions we can then take $\xi(X) = \pm\eta$.

Step 2. Now we need to find $P_0(X,0)$. If we multiply the first order equations by functions $-U_0$, V_0, P_0, respectively (which represent solution to the zeroth order equations with ξ replaced by $-\xi$), and integrate each equation over $Y \in (-\infty, 0)$, then we obtain after integration by parts and combining non-vanishing terms

$$\frac{\partial}{\partial X}\int_{-\infty}^{0}\left[\frac{i\omega\rho\xi}{\lambda^2}(V_0^2 - U_0^2) + U_0 P_0\right]dY = 0,$$

i.e.

$$\int_{-\infty}^{0}\left[\frac{i\omega\rho\xi}{\lambda^2}(V_0^2 - U_0^2) + U_0 P_0\right]dY = C_0,$$

where C_0 is a constant independent of X. Notice that all terms containing U_1, V_1 and P_1 have disappeared because of the vanishing factors in front of them. Now substitute U_0, V_0 and P_0 by their expressions in terms of $P(X,0)$, and integrate. We get

$$(P_0(X,0)^2) = \frac{2\omega\rho C_0(\sqrt{\xi^2 + i\lambda^2})^3\sqrt{\xi^2}}{\xi(\sqrt{\xi^2 + i\lambda^2} - \sqrt{\xi^2})(\sqrt{\xi^2 + i\lambda^2}\sqrt{\xi^2} - i\lambda^2)},$$

which gives us the solution.

3 Auditory Nerve (AN)

3.1 AN Structure

The auditory nerve is a collection of axons connecting the peripheral auditory system and the auditory areas of the brain. It is made up of approximately 30,000 to 55,000 nerve fibers, depending on species. About 95% of them are afferent, projecting from cell bodies in the cochlea to cochlear nuclei in the brainstem, and the rest are efferent, coming to cochlea from cells in the olivary complex (also part of the brainstem, see below). The afferent neurons are divided into Type I and Type II, based on their morphology: type I cells are large, have bipolar shape, their axons are large and myelinated, and, therefore, fast; type II cells are smaller, have different shape and non-myelinated axons. In addition, type I fibers innervate inner hair cells in many-to-one fashion, and type II fibers innervate many-to-many outer hair cells. Very little is known about the functional properties of type II afferents, partially because type I fibers are much easier for physiologist to record from, due to their large size and number. The role of efferent neurons in modifying the auditory input is also not yet clear. So, the rest of this text will focus on type I afferents.

3.2 Response Properties

Spontaneous Rates

In mammals, afferents can be divided into low, medium and high spontaneous rate fibers. The spontaneous rate (firing rate in the absence of stimuli) is determined by pattern of hair cell innervation, although the mechanisms are unclear. It could be that either smaller size of low spontaneous rate fibers makes them less excitable, or less transmitter is released from the hair cells to low spontaneous rate fibers. The variety of available fiber sensitivity provides a way of encoding wide range of intensities in auditory nerve (see *Intensity sensitivity* below).

Thresholds

The threshold is defined as minimal intensity of the stimulus that increases firing rate above the spontaneous level. Note that this is not a real threshold because, for example, for very low frequency fibers, sound first synchronizes spikes before there is ever an increase in rate. Usually the threshold is about 1 dB in mammals.

Latencies

Auditory fibers are also characterized by latencies in their responses. The latencies can be measured using brief stimuli, for example, clicks. Fibers tuned to different frequencies respond to a click at different latencies. Most of the delay originates from travel time of wave along basilar membrane. High frequency areas on the membrane are stimulated first, i.e. high frequency cells have shorter latencies.

Frequency Tuning

If we fix the intensity of the stimulus at any particular level and look at the firing rate of the given fiber at different frequencies, we find that the response function is usually single-peaked. This means that the cell has a preferred range of frequencies, inherited from its innervation of cochlea.

A typical way to characterize the basic properties of an auditory cell, is to probe its responses to a variety of *pure tones* (stimuli with only one frequency component and constant intensity). These responses are often summarized by marking the areas of intensity and frequency that produce a change in the firing rate of the cell (*response areas*). The border of this area is called a *tuning curve* of the cell. For auditory nerve fibers (Fig. 8) these areas usually have triangular shape, pointing down, and most of the area corresponds to an increase in firing rate (excitatory). The frequency at which the tip of the response area is located, i.e. the frequency at which the cell is most sensitive, is called the cell's characteristic frequency (CF) or best frequency (BF). The CF of each cochlear afferent is well-defined and it provides accurate information about the position on the cochlea of the hair cell that the afferent innervates. Approximately, the linear distance on the cochlea is proportional to the logarithm of CF. The shapes of the response areas change systematically with CF. The response areas of high-CF fibers have very steep high-frequency slope, and elongated tip (Fig. 8). Response areas of lower-CF afferents are relatively broader and more symmetrical. Notice that often relative bandwidth (normalized to CF), rather than absolute bandwidth, is used to characterize the sharpness of tuning; and relative bandwidth increases with CF. A common measure of frequency tuning is the Q_{10}, defined as the CF divided by the bandwidth at 10 dB above CF threshold. Q indexes also get higher with increasing CF.

Intensity Tuning

Next property of the auditory nerve fibers, that we will be discussing, is the intensity tuning. Let us keep the frequency constant and consider responses, in

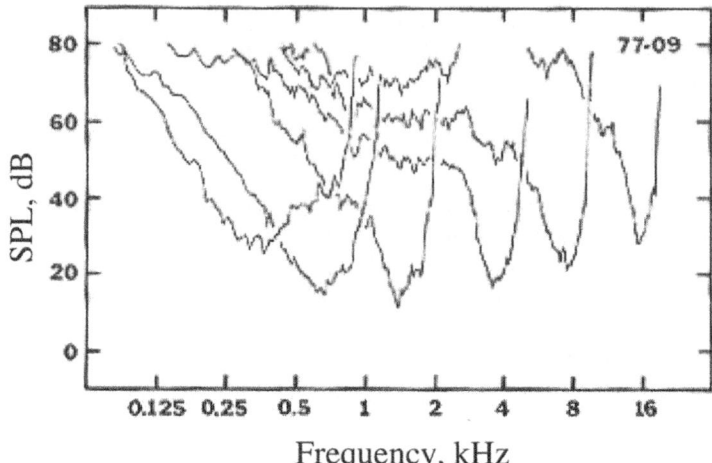

Fig. 8. Examples of frequency-threshold tuning curves for chinchilla auditory nerve fibers. Area above each curve is the response area of the given fiber. Reprinted by permission of VCH Publishers, Inc. from [86].

terms of firing rate, at different intensity levels (rate-intensity function). This function is generally monotonically increasing over 40-50 dB above threshold, and then saturates. The range over which this function is increasing is termed the *neuron's dynamic range*. Maximum dynamic range is usually at CF. Notice that the dynamic range of individual cells (40 dB) is 10^7 times smaller than the total range of human hearing (110 dB). To build such a wide range out of small-range elements, the auditory system makes use of the diversity in sensitivity of individual cells. The threshold levels of less sensitive fibers are situated at the upper end of the range of the more sensitive ones. In this way one set of neurons is just beginning to fire above their spontaneous rate when the other group is beginning to fire at their maximal rate. Additionally, at very high sound levels, the frequency tuning properties of the cochlea break down. Thus, we are able to tell that a 102 dB sound is louder than a 100 dB sound, but it is hard for us to determine whether it has a different frequency.

Phase Locking (Transmission of Timing Information)

Phase-locking of one (potentially stochastic) process with respect to another (often periodic) means that the events of the former preferentially occur at certain phases of the latter; in other words, that there is a constant phase shift between the two.

In the presence of a pure tone (periodic) stimulus, due to cochlear structure, inner hair cells are stimulated periodically by the pressure waves. If the frequency is low enough, then the depolarizations of the hair cell are also periodic and occur at certain phases of the stimulus cycle. This, in turn, generates a phase-locked release of neurotransmitter (that carries the neuronal signal to other cells) and, ultimately, leads to spikes in the auditory nerve that are also phase-locked to the stimulus cycle.

To observe or measure the phase-locking, people often build a histogram of phases of the recorded spikes within a period of the stimulus (Fig. 9). To have a meaningful resolution, given that stimuli frequencies can be quite high, bin widths have to be in the microsecond range. Phase-locking is equivalent to spikes accumulating at some parts of the periodogram, i.e. forming a peak.

A typical way to quantify phase-locking is by computing vector strength of the period-histogram [30]. Vector strength of a periodic function is the norm of the first complex Fourier component, normalized by the norm of the zeroth Fourier component. Normalization is included to remove dependence on the mean firing rate. This quantity can vary between 0 and 1, with 0 indicating low phase locking (for example, in case of uniform distribution of spikes), and 1 indicating high phase locking (when all spikes fall in the same bin). Figure 9 shows an example of periodograms and vector strengths of an auditory fiber in response to stimuli of different frequencies.

Phase-locking generally occurs for stimuli with frequencies up to 4-6 kHz. Even high frequency fibers, when stimulated with low frequencies at intensities 10-20 dB below their rate threshold, will show phase locking of spikes, without increase in firing rate [39]. Decline of phase-locking with increase in frequency originates in the parallel reduction of the AC component of the inner hair cell response.

Phase-locking plays an important role in our ability to localize sounds, particularly at low frequencies, at which interaction of sound waves with the body and the pinna is reduced. We are able to tell which direction a sound is coming from (its azimuth), based on time delay between when it reaches our right and left ears. The idea is that the sound, originating on one side of the head, must follow a longer path and consequently takes a longer time traveling to one ear than to the other. Then, due to the phase locking, the time delay is translated into the phase shift between spike trains, originating in the different sides of the brain. These delays are detected in some parts of the auditory pathway superior olivary complex, with precision as small as 20 microseconds. We will come back to the issue of phase-locking and sound localization when we talk about superior olivary complex below.

3.3 How Is AN Activity Used by Brain?

One of the basic questions one can ask about how the AN signal is decoded is the following: how does a neuron differentiate between change in rate due to change in intensity vs change in rate due to change in frequency? While the complete answer is not yet clear, it seems that the key is to look not at the responses of the single cell, but at the distribution of the responses across a population. The basic view is that the loudness can be carried by the total level of the population activity, and the frequency — by the position of the

UNIT 68·II6·6

94 dB SPL

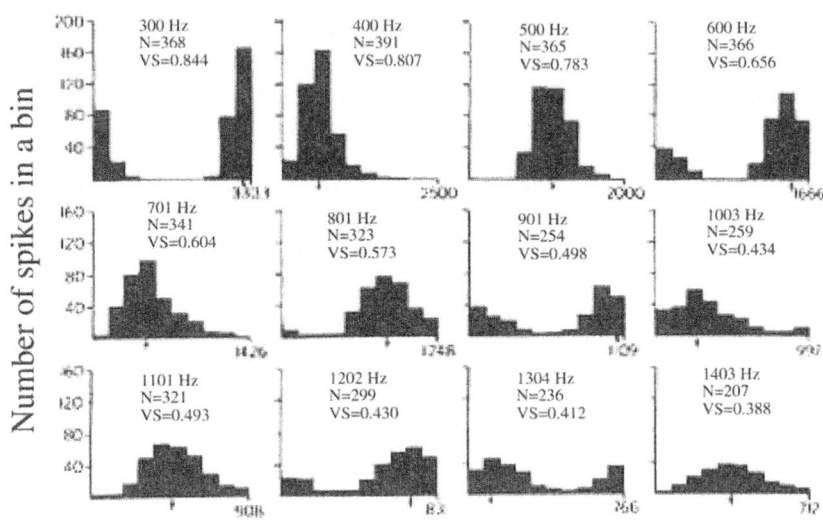

One period of stimulus time, msec

Fig. 9. Periodograms of responses of an auditory nerve fiber. Each panel shows response to one stimulus frequency, and horizontal axis spans one stimulus period. Above the histogram is the stimulus frequency, total number of spikes (N) and the vector strength (VS). Reprinted by permission of Acoustical Society of America from [5].

activity peak within the population (place principle) and/or by the temporal structure of the spike trains (using phase-locking — volley principle).

In this simple view, one considers complex sounds as made up of sinusoids (Fourier components). This linear approximation works well (at normal acoustic pressures) for the outer ear, less well for middle ear and only partially for the inner ear. The inner ear, anatomically, breaks sound into bands of frequencies, that persist for many stages of processing. How the information from these different bands is put back together is unknown.

In reality things are not linear. We will give two examples of non-linearities. The first example shows that the responses of AN fibers to tones depend on the spectral and temporal context. We illustrate this by describing a two-tone suppression: response of AN fiber to a given tone can be suppressed by a presence of another tone of a given frequency. The second example illustrates the non-linearity in a phenomenon of "hearing the missing fundamental": tone combinations can create perception of frequencies that are not really present in the signal. Now we describe these examples in more detail.

Example 1. *Two-tone suppression* occurs in recordings from cochlea or afferents. It is defined as reduction in response to one tone in presence of a second tone. It depends upon levels and frequencies of the two tones. To illustrate this phenomenon, one can map response area of a cell (Fig. 10), then choose a test tone near CF (triangle in Fig. 10), and mark by shading the tonal stimuli that suppress responses to the test tone. Regions of overlap of the shading with the response area show that some tones that are excitatory by themselves, can suppress responses to CF tones. Latency for suppression is as short as initial response, i.e. it is not caused by efferents. It is shown that major component of the suppression originates from basilar membrane motion.

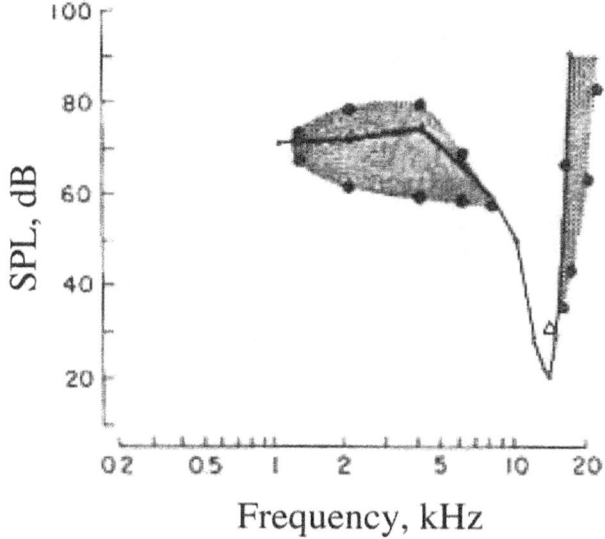

Fig. 10. Example of two-tone suppression. Tuning curve (black line), showing the lower border of the response area of an auditory nerve fiber. Response to test tone (triangle) can be suppressed by any of the tones in the shaded area. Recording from chinchilla auditory nerve fiber, reprinted from [33] with permission from Elsevier.

Example 2. The second example of non-linearity is that in some specific types of experiments it is possible to hear a sound of certain frequency even though the ear is being stimulated by several other frequencies. This phenomenon is sometimes referred to as *"hearing the missing fundamental"*. For example, three frequencies are played simultaneously F1=4000 Hz, F2=6000 Hz, F3=8000 Hz. Rather than hearing these three frequencies, the listener actually hears a 2000 Hz sound. It has been shown that the region of the cochlea with its threshold tuned to 2000 Hz do not fire any faster, while the neurons "tuned" to 4000 Hz, 6000 Hz and 8000 Hz are firing well above their spontaneous rate. To account for this paradox some scientist have proposed that frequency discrimination is not done based on basilar membrane resonance, but on timing information. Due to phase locking and because all frequencies present are multiples of 2000 Hz, multiple auditory nerve fibers will fire simultaneously at every cycle of the 2000 Hz oscillation. It is hypothesized that if only this timing information is used to interpolate the frequency content of the sound, the brain can be tricked into hearing the missing fundamental.

3.4 Modeling of the Auditory Nerve

Mathematical modeling of the responses of the auditory nerve fibers has mostly been phenomenological. The models aimed to described as accurately as possible the experimentally known characteristics of fibers' responses. For example, to match the responses to single simple tones or clicks and to temporal combinations of different stimuli [16, 104, 119], or to determine the lower thresholds of the single-pulse responses [111]. Prevalence of this type of modeling is due to the fact that the biophysics of the hair cells, where auditory nerve fibers originate, has not been worked out in sufficient detail to allow construction of more biophysically realistic models. However, the auditory nerve fiber modeling has recently become one of the examples of how just how much theoretical studies can contribute to the field of neuroscience. Recent models of Heinz et al. [34, 35] provide a link between auditory fiber physiology and human psychophysics. They find how results equivalent to human psychophysical performance can be extracted from accurate models of the fiber responses. In essence, this demonstrates the "code" in auditory nerve fibers via which the auditory system extracts all necessary information.

4 Cochlear Nuclei

As fibers of the auditory nerve enter the cochlear nuclei (CN), they branch to form multiple parallel representations of the environment. This allows parallel computations of different sound features. For example, sound localization and identification of a sound are performed in parallel. This separation of processing pathways exists both in the CN and further upstream in the brainstem.

Some information carried by auditory nerve fibers is relayed with great fidelity by specific neurons over specific pathways to higher centers of the brain. Other CN neurons modify the incoming spike trains substantially, with only certain elements of the input signal extracted prior to transmission to the next station in the auditory pathway.

One of the features that is inherited in CN from the auditory nerve is tonotopy (continuous variation of the characteristic frequency with the position of the cell, as on basilar membrane). But, interestingly, tonotopic projections are not point-to-point. Each characteristic frequency point on basilar membrane projects to an iso-frequency *plane* across the extent of the cochlear nucleus. Thus cochlear place representation is expanded into a second dimension in brain. These tonotopic sheets are preserved in projections all the way to cortex. We contrast this with the visual and somatosensory systems, where maps reflect a location of stimulus in space and the representations are point to point.

In this section we will first describe general structure of the CN and outline its input and output streams. Then we will describe in more details the properties of cells that make information processing in the CN possible.

4.1 Basic Features of the CN Structure

As we mentioned above, there are parallel streams of information processing in the cochlear nuclei. Thus, the CN is naturally subdivided into areas that house cells with different specialized properties, receive different patterns of inputs and project to different targets. The main subdivision structures are the ventral cochlear nucleus (VCN) and the dorsal cochlear nucleus (DCN) (Fig. 11).

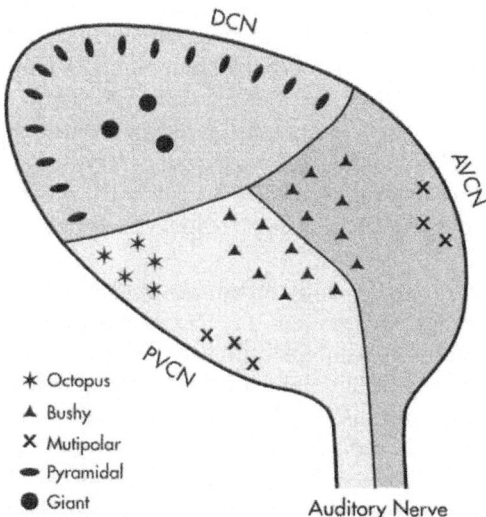

Fig. 11. Schematic of cell type location in CN. Main parts of the CN are marked by color (green is dorsal cochlear nucleus (DCN), yellow is posteroventral cochlear nucleus (PVCN) and blue is anteroventral cochlear nucleus (AVCN)). Cell types are marked by symbols. Illustration is made by Brook L. Johnson.

The VCN contains four types of principal cells: globular bushy cells, spherical bushy cells, multipolar cells and octopus cells. We will describe their properties below, and just mention now that they play a very important role in processing different sorts of timing information about the stimuli.

Unlike VCN, DCN is layered and has interneurons. The outermost layer, called the superficial or molecular layer, contains cell bodies and axons of several types of small interneurons. The second layer, called he pyramidal cell layer, has the cell bodies of pyramidal cells, the most numerous of the DCN cell type, and cartwheel and granule cells. The deep layer contains the axons of auditory nerve fibers as well as giant cells and vertical cells.

4.2 Innervation by the Auditory Nerve Fibers

The main external input to the CN comes from the auditory nerve fibers. There is also input to granule cells that brings multimodal information from widespread regions of the brain, including somatosensory, vestibular and motor regions, but we will not consider it here.

As a reminder, in mammals there are two types of the auditory nerve fibers. Type I fibers innervate one inner hair cell each, the fibers are thick and myelinated, and they constitute 90-95% of all fibers. Type II fibers innervate outer hairs cells, they are thinner and unmyelinated.

Type I fibers in the cochlear nuclei form two branches: the ascending branch goes to the anteroventral region of the cochlear nucleus (AVCN) and the descending branch to the posteroventral region (PVCN) and parts of DCN. Type II fibers project to DCN, but because it is not clear if they carry auditory information, type II fibers are, once again, not considered here.

Auditory nerve fibers make synapses to all cell types in the cochlear nuclei, except in the molecular layer of the DCN and in the granule cells region. The innervation is tonotopic within each principal cell type in the VCN and in the deep layer of DCN.

Nerves form different types of terminals onto different cell types and/or different cochlear nucleus divisions. The terminals range from small to large endbulbs. The largest are the endbulbs of Held (*calices of Held*) onto the bushy cells. Each such terminal contains hundreds of synapses. This allows to inject a lot of current into postsynaptic cell every time the pre-synaptic signal arrives. Perhaps only one endbulb is needed to fire a cell, and the transmission through these connections is very fast and reliable. Other CN cell types receive input from auditory nerves through more varicose or bouton-like terminals, located at their dendrites. In these cases more integration is required to fire a cell.

All auditory nerve synapses use glutamate as the neurotransmitter, and are depolarizing (excitatory). Often postsynaptic cells have specialized "fast" receptors (of AMPA type), important in mediating precise coding. Interestingly, in young animals glutamate receptors in the VCN are dominated by very slow NMDA receptors and are replaced by AMPA receptors in the course of development. The reason for this change is not presently clear.

Another important feature that is modified with age is synaptic plasticity. In many neuronal systems the strength of synapses are modified, both on the short (during a single stimulus response) and on the long time scale. In contrast, synaptic transmission by adult auditory nerve fibers shows little plasticity. Synaptic depression (decrease in synaptic efficacy on time scale of 100 msec) is prominent in young animals and decreases with age.

4.3 Main CN Output Targets

Figure 12 shows the two major fiber bundles that leave the cochlear nucleus in relation to other structures in the brainstem. It also shows some of the auditory neuronal circuits of the brainstem. At the output of the cochlear nucleus, spherical and globular bushy cells (SBC and GBC) project to the medial (MSO) and lateral (LSO) superior olivary nuclei and the trapezoid body (MNTB and LNTB). The MSO and LSO compare input from the two ears and perform the initial computations necessary for sound localization in the horizontal plane (see next section). Excitatory inputs to the superior olive come mostly from the spherical bushy cells and inhibitory inputs come from the globular bushy cells via the inhibitory interneurons in the trapezoid body.

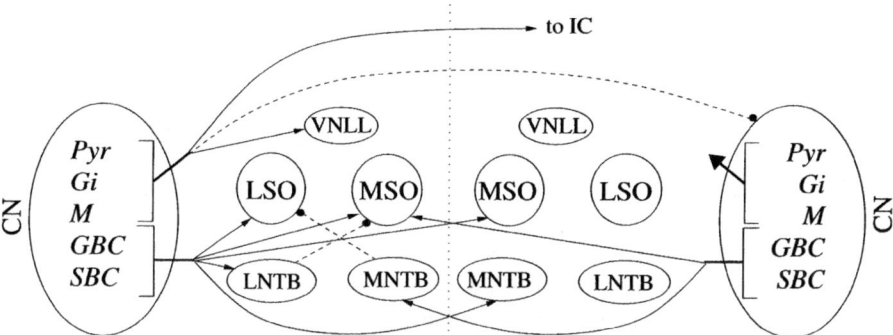

Fig. 12. Main ascending connections from cochlear nucleus (CN). Main CN cell types: globular bushy cells (GBC), spherical bushy cells (SBC), multipolar (M), giant (Gi), pyramidal (Pyr) make connections to medial nucleus of the trapezoidal body (MNTB), lateral nucleus of the trapezoidal body (LNTB), medial superior olive (MSO), lateral superior olive (LSO), ventral nucleus of lateral lemniscus (VNLL)and inferior colliculus (IC). All connections are symmetric with respect to midline (dotted line), but not all are shown. Excitatory connections are shown with solid lines and inhibitory with dashed.

Multipolar, giant, and pyramidal cells project directly to the inferior colliculus, where all ascending auditory pathways converge (see section 6). The octopus cells project to the superior paraolivary nucleus and to the ventral nucleus of the lateral lemniscus (VNLL). The granule cells axons project to the molecular layer of the DCN, where they form parallel fibers. These parallel fibers run orthogonal to the auditory nerve fibers and cross isofrequency layers. The input from these two sources is combined by the principal cells of the DCN – the pyramidal cells.

4.4 Classifications of Cells in the CN

In the current prevalent view, each cochlear nucleus cell type corresponds to a unique pattern of response to sound; this is consistent with the idea that each type is involved in a different aspect of the analysis of the information in the auditory nerve. The diversity of these patterns can be accounted for by three features that vary among the principal cell types: (1) the pattern of the innervation of the cell by ANFs, (2) the electrical properties of the cells, and (3) the interneuronal circuitry associated with the cell.

To study the correlation between physiological and anatomical properties of the cells experimentally, it is necessary to make intracellular recordings of responses to various current injections and then fill the cells with a dye to image their shape. These experiments are hard to perform and while there seems to be evidence that the correlation between structure and function in these cells is strong, this evidence is not conclusive.

Physiological Types

Three major types of VCN cells (bushy, multipolar and octopus) probably correspond to different physiological types: primary-like, chopper and onset.

Primary-like (spherical bushy) and primary-like with notch (globular bushy) responses are very much like auditory nerve. Initial high burst of spikes (>1,000 Hz) followed by decline to maximum rate of no more than 250 Hz. Pause (notch) is mostly due to refractory period of cell.

Choppers are major response type in PVCN, also found in other divisions. In response to high frequency tones, they discharge regularly, independent of stimulus frequency and phase with firing rates sustained at up to 200-500 Hz. These appear to be multipolar cells.

Onset responses have spikes at the onset of stimulus, and then fewer or none. Standard deviation of first spike latency is very small, about 100 μsec. Those few spikes that are produced during the ongoing stimulus phase lock to low frequency tones. These cells are located mostly in PVCN. Some of them are octopus cells and others are large multipolars. The properties of these responses (precise onset and brief transient activation) allow them to play an important role in temporal coding.

DCN cells exhibit wide variety of response types. Their functions probably are related to pinna movement and echo suppression (which allows you to not hear yourself many times when you speak in a small room). In addition, there is also somatosensory input through granule cells that has to be combined with the auditory one.

In Vitro Physiology

To complicate things further, slice physiology in VCN provides additional view of types. In these studies intracellular recordings are made from isolated cells. It is possible to either control the current flowing across the cell membrane and measure the induced voltage (current clamp), or to fix the membrane voltage and to record the resulting trans-membrane current (voltage clamp). The recordings reveal two major physiological response types. Type 1 fires regular train of action potentials in response to depolarization. Also, it has relatively linear current-voltage curve. Type 2 fires just one, or a few, action potentials in response to ongoing depolarization. In addition it has very nonlinear current-voltage curve, with zero (reversal potential) at about -70 mV and much steeper slope in depolarizing than in hyperpolarizing direction.

Differences between types 1 and 2 are largely due to change in potassium currents. Type 2 cells have low threshold potassium current that is partially activates near rest values of voltage and strongly activates when voltage rises, repolarizing the membrane and halting the response. Many type 1 cells are multipolar and many type 2 cells are bushy (see more on this below, under Bushy and Multipolar cells).

4.5 Properties of Main Cell Types

Bushy Cells

Bushy cells have short (<200 μm), bushy dendritic trees. Their synaptic input is located mainly on the soma, with few synapses on their dendrites. Two subtypes of bushy cells are recognized as spherical and globular.

Bushy cells receive inputs from auditory nerve through large synaptic terminals (calyx of Held), have primary-like responses and accurate temporal coding. As mentioned earlier, spherical bushy cells and globular bushy cells differ in their locations and their projections. Spherical cells are located in the more anterior region, and project to MSO, while globular cells project to LSO and trapezoid body. Spherical cells have one or a few short dendrites that terminate in a dense bush-like structure near the soma, and globulars have more ovoid somata and larger, more diffuse dendritic trees. Spherical cells also have lower characteristic frequencies, better phase-locking and are specialized for accurate encoding of AN signal. Globular bushy cells sometimes chop or have onset responses, receive more and smaller endbulbs, and have higher characteristic frequencies. Probably these cells started out as one type and have diverged with evolution of sensitivity to higher frequencies.

Bushy cells respond to sound with well-timed action potentials. The characteristics of both the synaptic current and the postsynaptic cell properties are made so that they shape the timing of the response.

Responses to Sounds

Figure 13A shows typical responses to tones of bushy cells. Large unitary synaptic events (excitatory post-synaptic potentials; EPSPs) from between one and three endbulbs cause spherical bushy cells to fire whenever the auditory nerve fibers do (except when the cell is refractory). This one-spike-in, one-spike-out mode of processing means that the responses to sound of spherical bushy cells resemble those of auditory nerve fibers, and for this reason they are called "primary-like". Evidence that primary-like responses reflect a "one-spike-in, one-spike-out" is provided by their action potential (AP) shapes. The AP is preceded by a pre-potential (reflecting the build up of the endbulb activity) which are almost always followed by the postsynaptic component of the spike, demonstrating security of the synapse.

Globular bushy cells give similar response, primary-like-with-notch. They differ from primary in that in the beginning of the response there is a precisely timed peak followed by a notch. Precise timing is helped by the convergence of large number of fibers. If a globular cell needs one input to fire, then there is a very high probability that one of the incoming fibers will activate early on and cause a spike. This initial peak will be followed by recovery period (refractoriness), generating notch.

Electrical Characteristics

As a response to injection of constant depolarizing current bushy cells produce one to three spikes at the onset and then settle to a slightly depolarized constant voltage value. This is due to the low-threshold potassium current. It is partially activated at rest, then during spiking it strongly activates increasing membrane conductance and thereby shortening the membrane time constant and repolarizing the cell. To induce firing, the input must be very strong and fast. The short time constant blocks temporal integration of inputs. Rapid temporal processing permits bushy cells to preserve information about the stimulus waveform information that is necessary for sound localization.

Multipolar Cells

Multipolar cells have multiple, long dendrites that extend away from the soma in several directions. These cells are also sometimes referred to as "stellate". Two major classes of multipolar cells have been described. The cells of one group, T-multipolars (planar), have a stellate morphology with dendrites aligned with auditory nerve fibers , suggesting that these cells receive input from a restricted range of best frequencies. Their axons project through the trapezoid body (hence the "T") to the contralateral IC. Cells of the second group, D-multipolars (radiate), have dendritic fields that are not aligned with auditory nerve fibers . Their axons project to the contralateral cochlear nucleus.

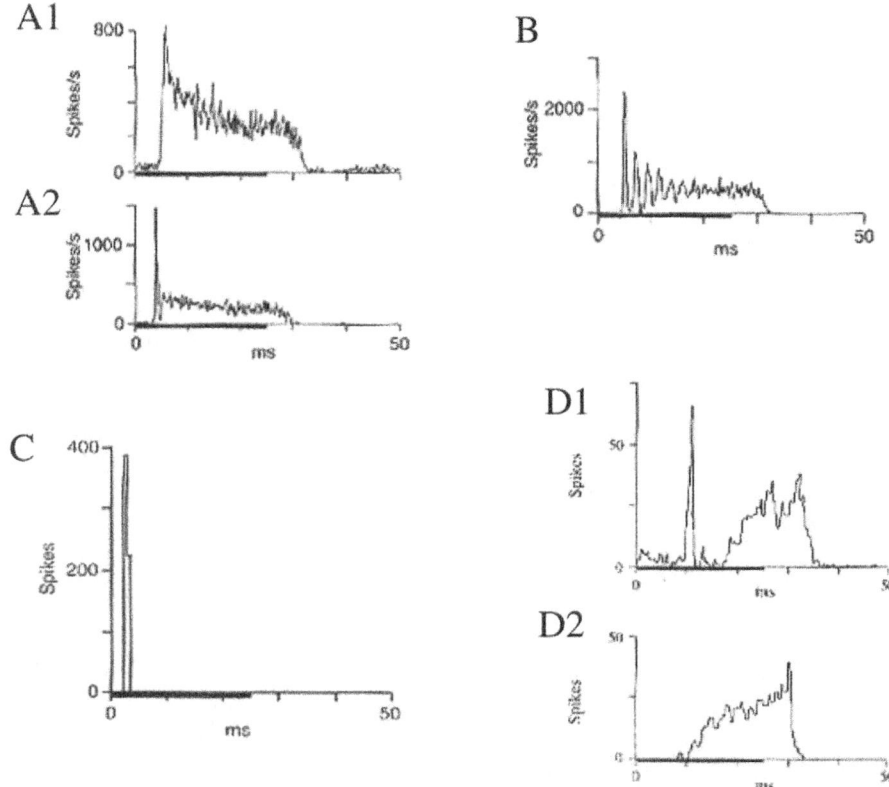

Fig. 13. Different response types. *A1:* Primary-like response (bushy cell). *A2:* Primary-like with notch response (bushy cell). *B:* Chopper response from T-multipolar cell. *C:* Single-spike onset response. *D1:* Pauser response. *D2:* builder response. In all panels dark black line under the axes is the tone presentation. Panels A and B are used by permission of The American Physiological Society from [8]. Panel C is reprinted from [42] with permission from Elsevier. Panel D is reprinted by permission of VCH Publishers, Inc. from [29].

Both T- and D-multipolar cells in the VCN serve the role of interneurons through their axon collaterals. These terminate locally within VCN as well as projecting to the deep DCN. T-Multipolars are excitatory and D-multipolars are inhibitory and glycin-ergic.

Responses to Sounds

Figures 13B shows responses characteristic of T-multipolar cells. T-multipolar cells respond to tones by firing at regular intervals independent of frequency of the tone, a pattern called chopping. The reproducibility of firing gives histograms of responses to sound a series of characteristic modes that is independent of the fine structure of the sound. The intervals between nodes are of equal duration and correspond to the intervals between spikes, with one spike per node. The peaks are large at the onset of the response because the latency to the first spike is quite reproducible in chopper neurons; the peaks fade away over the first 20 msec of the response, as small variations in interspike interval accumulate and spike times in successive stimulus repetitions diverge. The chopping firing pattern must arise from the intrinsic properties of the cells themselves, since it does not reflect properties of the inputs. Moreover, the same pattern is elicited by depolarization with steady currents (Fig. 14).

D-multipolar cells are broadly tuned and respond best to stimuli like noise but only weakly to tones. This is due to the fact that D-multipolars receive input from many auditory nerve fibers on their somata and on dendrites that spread across the frequencies. D-multipolars also respond with a precisely timed onset spike to tones, but (unlike octopus cells) they give some steady discharge after the onset spike. Also, unlike octopus cells, the firing pattern in response to tones is shaped by inhibition along with the intrinsic electrical properties.

Electrical Characteristics

Unlike bushy cells, multipolars are capable of temporal summation of successive EPSPs. Recall from above, that bushy and multipolar cells have AMPA receptors with similar rapid kinetics. The differences between unitary voltage responses arise because the decay of the unitary event, and therefore the degree of temporal integration, is determined by the membrane time constant of the cell. For bushy cells this is short (2-4 msec), whereas for multipolar cells it is longer (5-10 msec).

Octopus Cells

Octopus cells in VCN (Fig. 11) are contacted by short collaterals of large numbers of auditory nerve fibers through small terminal boutons. Octopus cell dendrites are oriented in one direction, inspiring their name. The orientation is perpendicular to the auditory nerve fibers so that the cell bodies encounter fibers with the lowest best frequencies, and the long dendrites extend toward fibers that encode higher frequencies. Each cell receives input from roughly one-third of the tonotopic range, i.e. the frequency tuning is broad and contribution of each fiber to the octopus cell response is very small. Also EPSPs that an octopus cell receives are very brief, between 1 and 2 msec in duration.

Responses to Sounds

Octopus cells behave like bushy cells in that the low input resistance prevents temporal summation of inputs. But instead of receiving a few large inputs, octopus cells receive small inputs from many fibers. Thus, many fibers have to fire simultaneously to drive an octopus cell.

Such synchronous firing is not achieved with ongoing pure tone stimuli. Rather, it happens at stimulus transients. For example, at the onset of a pure tone (thus the term onset response, Fig. 13C), or at a rapid fluctuation of a broadband stimulus such as at the onset of a syllable or during a train of clicks. In fact an octopus cell can faithfully follow a train of clicks with rates up to 500 Hz.

Electrical Characteristics

In the case of octopus cells, the fast membrane time constant or high membrane conductance is generated by presence of two opposing voltage-sensitive currents. One of them activates with hyperpolarization and has a reversal potential near -40 mV, the other is a potassium current (reversal potential near -80 mV) that activates with depolarization. At rest both currents are partially activated. In fact, they are quite large, but they compensate one another. In addition, the voltage-gated conductances of these currents are very sensitive to voltage near resting potential. This means that any voltage fluctuation activates these currents which counter the voltage change. As a result, EPSPs are always small and brief and the cell is only sensitive to synchronous inputs. Moreover, any depolarizing current that rises too slowly is bound to activate the potassium conductance and to preclude the cell from responding. This mechanism makes octopus cells sensitive to the rate of rise of their input, not its amplitude. In a sense they are sensors of the derivative of the input.

Pyramidal Cells

Pyramidal (also called "fusiform") neurons in the DCN are bipolar, with a spiny dendritic tree in the molecular layer and a smooth dendritic tree in the deep layer. The cell bodies of pyramidal cells form a band in the pyramidal cell layer. The smooth dendrites are flattened in the plane of the isofrequency sheets, where they receive input from the auditory nerve. The spiny dendrites span the molecular layer and are contacted by parallel fibers at the spines.

Responses to Sounds

Pyramidal neurons in the DCN show pauser and buildup responses to sound. The examples shown in Fig. 13D are typical of anesthetized animals, where the inhibitory circuits of the DCN are weakened. The response shows a poorly timed, long latency onset spike followed by a prominent pause or a slow buildup in response with a long latency.

Electrical Properties

The pauser and buildup characteristics seem to derive from a transient potassium conductance. This conductance is inactivated at rest. If a sufficient depolarizing stimulus arrives, the cell simply responds with a spike. However, if the cell is initially hyperpolarized, this removes the potassium current inactivation. Any subsequent depolarization will activate the potassium current transiently, producing the long latency of a pauser or buildup response. *In vivo* a sufficient hyperpolarization isprovided by inhibitory synaptic inputs as an aftereffect of a strong response to an acoustic stimulus.

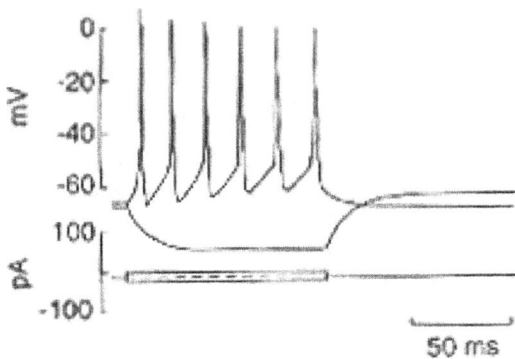

Fig. 14. Responses of T-multipolar cell to intracellular injections of depolarizing and hyperpolarizing currents. Membrane potential (top) and current time courses (bottom). Spiking is in response to the depolarizing current. Reproduced with permission from [59], Copyright 1991 by the Society for Neuroscience.

Some of the Other Cell Types

Giant cells are large multipolar cells located in the deep layers of the DCN. They have large, sparsely branching, dendritic trees that cross isofrequency sheets. Giant-cell axons project to the contralateral inferior colliculus.

Granule cells are microneurons whose axons, the parallel fibers, provide a major excitatory input to DCN through the molecular layer. Granule-cell axons terminate on spines of the dendrites of pyramidal cells, on spines of cartwheel cells, and on the stellate cells.

The vertical cells are inhibitory interneurons that project to their isofrequency sheets in both DCN and VCN; they inhibit all of the principal cells in the cochlear nucleus, except the octopus cells. Vertical cells are narrowly tuned, and they respond most strongly to tones at a frequency near their characteristic frequency.

Cartwheel cells are inhibitory interneurons whose numerous cell bodies lie in the pyramidal cell layer of the DCN. Their dendrites span the molecular layer and are densely covered with spines that are contacted by parallel fibers. They contact pyramidal, giant, and other cartwheel cells through glycinergic synapses. The reversal potential of the cartwheel-to-cartwheel cell synapses lies a few millivolts above the resting potential and below the threshold for firing so that its effect is depolarizing for the cell at rest but hyperpolarizing when the cell has been depolarized by other inputs. Thus, the effects of the cartwheel cell network are likely to be context-dependent: excitatory when the cells are at rest but stabilizing when the cells are excited.

Figure 15 shows responses to sound of cartwheel cells. Cartwheel cells are the only cells in the cochlear nucleus with complex action potentials, which reflect a combined calcium and sodium spike. Many of the cartwheel cells respond weakly to sounds and no particular pattern of response is consistently observed. They probably mainly transmit non-auditory information to the principal cells of the DCN.

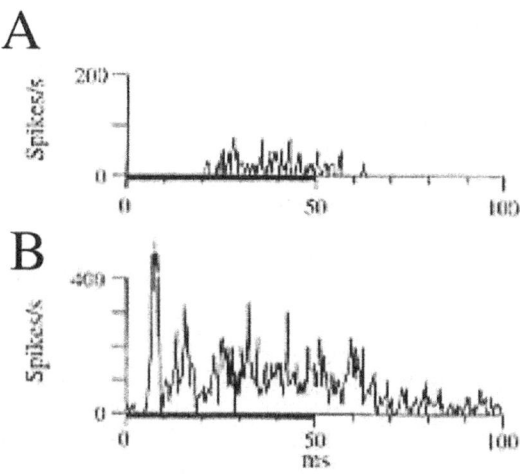

Fig. 15. Weak (A) and strong (B) responses of cartwheel cells. Used by permission of The American Physiological Society from [71].

As a side note: cartwheel cells share many features with cerebellar Purkinje cells. For example, both of these fire complex action potentials; and genetic mutations affect Purkinje cells and cartwheel cells similarly.

4.6 Modeling of the Cochlear Nuclei

As data on the membrane properties of cochlear nucleus neurons have accumulated, it has become possible to use biophysical models of Hodgkin-Huxley type to explore the different behaviors described in the previous section. These models give researchers ability to test whether the biophysical mechanisms discussed above can indeed underlie the observed cellular responses. In particular, when many ionic currents have been identified in a given cell type, a model helps identify which of them are the key ingredients in shaping each feature of the cell's behavior.

For example, a model of bushy cells [85] demonstrates that the presence of a low-threshold potassium current in the soma of the cell can account for responses of both spherical and globular bushy cells depending on number and strength of inputs. When inputs are few and large the response matches characteristics of the primary-like response, but with larger number of inputs it becomes primary-like with notch. Another computational study of globular bushy cells [47] uses a much simpler underlying mathematical model (integrate-and-fire rather than Hodgkin-Huxley type). This allows to study in generality what increases or decreases synchrony in the output vs. the input, using globular bushy cell as an example.

Some of the other models have dealt in detail with responses of various cells in the dorsal cochlear nucleus [9, 80].

Another example of mathematical modeling in cochlear nucleus is a model of multipolar cell [6]. This is a more complex multi-compartmental model, which demonstrates which combination of known somatic and dendritic currents can accurately reproduce properties of a chopper response (regular firing with irregular or constant inputs).

5 Superior Olive. Sound Localization, Jeffress Model

The superior olive is a cellular complex in the brainstem of about 4 mm long. The cytoarchitecture of this complex defines three parts: the medial superior olive, the lateral superior olive, and the nucleus of the trapezoid body. It should be emphasized that the trapezoid body itself is not a nucleus; it is a bundle of fibers. The entire superior olive complex is surrounded by small cellular groups known as the preolivary or periolivary nuclei. The olivary nuclear complex is the first level in the auditory system where binaural integration of auditory signals occur. It is the key station to performing many of the binaural computations, including, in large part, localization of sounds.

Two nuclei within superior olivary complex that have been most extensively studied are the medial and lateral superior olivary nuclei (MSO and LSO). Other nuclei, such as medial and lateral nuclei of the trapezoid body (MNTB and LNTB) are named according to their position with respect to the MSO and LSO. There is also a group of cells called olivacochlear neurons. They project to the cochlea. These neurons can be activated by sound, and cause suppression of spontaneous and tone evoked activity in auditory nerves.

5.1 Medial Nucleus of the Trapezoid Body (MNTB)

The cell body of neurons in the MNTB receives input from axonal projection of cells in the contralateral VCN, primarily from the globular bushy cells (GBC). The GBC-to-MNTB synapses are large, endbulb type, i.e. they also provide reliable synaptic connection. For that reason the responses of many MNTB cells are similar to their primary excitatory input, the globular bushy cells.

Some cells with chopper type responses are also found. Because convergence of the bilateral auditory information has not occurred yet at the level of MNTB, the neurons there are responsive exclusively to sounds presented to the ear contralateral to the nucleus itself. As usual, each cell responds best to a characteristic frequency. They then send a short axon to the LSO where it forms inhibitory glycine-ergic synapses.

5.2 Lateral Superior Olivary Nucleus (LSO)

Principal cells receive inputs onto soma and proximal dendrites. Excitatory input is received directly from the ipsilateral VCN. Inhibitory input is from the contralateral VCN, transmitted through the MNTB. Even though there is an extra synapse in the path of the inhibitory signal, inputs from each ear arrive at the LSO simultaneously. This is possible due to the large reliable synapses associated with MNTB. The cells that receive excitatory input from ipsilateral ear and inhibitory input from the other ear are often called IE cells.

Because of the organization of their inputs, LSO cells are excited by the sound which is louder in the ipsilateral ear and softer in the contralateral ear. There is almost no response when a sound is louder in the contralateral ear than in the ipsilateral ear. As a result the sensitivity functions of LSO neurons to the interaural level difference (ILD) are sigmoidal. There are successful computational models by Michael Reed and his colleagues (e.g. [81]) that demonstrate how a spectrum of sound ILDs can be coded by a population of the sigmoidally-tuned LSO cells.

The characteristic frequencies of LSO cells are predominantly in the high frequency range complimenting the range of frequencies over which the MSO is responsive. The frequency tuning is sharp, as narrow as in the cochlear nucleus and the auditory nerve, because ipsilateral and contralateral inputs are well matched in frequency. However, ipsilateral (excitatory) inputs are slightly broader tuned. Therefore, to maintain a response level of an LSO neuron, as contralateral sound moves away from characteristic frequency it must become louder to continue counteracting the ipsilateral signal.

5.3 Medial Superior Olivary Nucleus (MSO)

The cell bodies of neurons in the MSO receive input from two sets of dendrites. One projects laterally from the cell and gets its input from the ipsilateral VCN. The other projects medially from the cell and receives its input from the contralateral VCN. The cells in the MSO, as in LSO, are classified according to their response to these two sets of dendritic inputs. If a cell is excited by both contralateral and ipsilateral input, it is classified as an EE cell. If a cell is excited contralateral, but inhibited by ipsilateral input, it is classified as an EI cell. If a cell is excited by ipsilateral, but inhibited by contralateral input, it is classified as an IE cell. If a cell is inhibited by both contralateral and ipsilateral input, that is it fires below its spontaneous rate when either ear is presented with an appropriate stimulus, it is classified as an II cell.

MSO neurons respond to both binaural and monaural stimuli. The monaural response however will always be submaximal when compared to similar binaural stimuli. The response tends to be frequency selective and the best frequency is typically the same for both ipsilateral and contralateral input in a given cell. The characteristic frequencies in the MSO tend to be in the low end of the hearing spectrum. In addition to a characteristic frequency, many neurons in the MSO respond best to a specific delay between ipsilateral and contralateral stimulus termed the "characteristic delay". They also have a "least characteristic delay" to which they respond worse than to monaural input. If the characteristic delay is preserved across frequencies or modulation frequencies, the cell is termed to be of "peak-type". If the least characteristic delay is preserved across frequencies the cell is termed "trough-type". Most of the cells in the MSO are of peak-type, and they form the basis for a famous Jeffress theory, which is described below.

5.4 Sound Localization. Coincidence Detector Model

Many animals, especially those that function in darkness, rely on sound localization for finding their prey, avoiding predators, and locating members of their own species. Therefore, it is not surprising that many mechanisms for sound localization, well-suited for different circumstances, have been created in the course of evolution.

Primary cues for sound localization are binaural. They include interaural differences in sound pressure level, time of arrival and frequency spectrum of the sound. It is also possible to localize sounds monaurally, but it is much less efficient (see, e.g., [106]). Historically, the most studied binaural cues have been interaural time and sound pressure level differences. Interaural time difference (ITD) arises because the sound, originating on one side of the head, must follow a longer path and consequently takes a longer time traveling to the distal than proximal ear (Fig. 16). The interaural level difference (ILD) results from the interference of the head and ears of the listener with the propagation of the sound so that a "sound shadow" may develop, i.e., the sound may be attenuated on the far side relative to the near one. In natural settings an acoustic signal usually gives rise to both time and amplitude differences. Under certain conditions only one of the cues retains sufficient magnitude to serve the listener. In particular, according to the duplex theory [79], interaural level differences are the primary cue for localizing high-frequency tones, whereas interaural time differences are responsible for the localization of the low frequency tones. The basis for this distinction is that for high-frequency sounds the wavelength is small compared with the size of the head and the head becomes an obstacle. At lower frequency sounds the wave length is longer and the head is acoustically transparent.

In recent years many departures from the duplex theory have been found. For example, for complex high-frequency signals ITD between low-frequency envelopes can serve as a localization cue [115]; direct physical measurement of ILD at the ears of the listener (e.g., in cats [108] and humans [21] show that ILD does not vary in a systematic fashion with azimuth; further, the interaural differences can be altered by factors such as head and pinnae shape; finally, interaural spectral differences, that develop for complex acoustic signals (particularly with high frequency components), provide considerable directional information that can enhance the localization task.

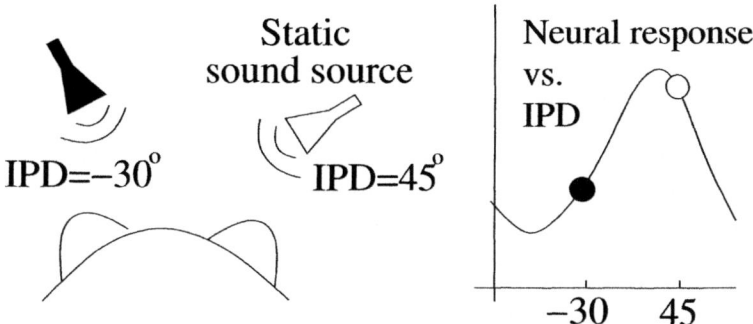

Fig. 16. When a sound source is placed not directly in front of the animal, but off-center, it generates ITD (and IPD). Two schematic sound source locations are shown in the left panel. If the response of a neuron is recorded at the same time, then it can be plotted vs. corresponding IPD (schematically shown in the right panel). This plot of response vs. IPD is the static tuning curve of the neuron.

Nevertheless, time and level differences continue to be considered as primary sound localization cues. Many species are very acutely tuned to them. Humans, for example, can readily sense interaural disparities of only a few decibels or only tens of microseconds [79, 102]. The sensitivity is even higher, for example, in some avians [46]. Besides psychophysical evidence, there is a great amount of physiological data, showing that in various structures along the auditory pathway (medial and lateral superior olive (MSO and LSO), dorsal nucleus of lateral lemniscus (DNLL), inferior colliculus (IC), auditory cortex) low-frequency neurons are sensitive to ITDs and high-frequency neurons to ILDs. What frequency is "low" or "high" varies with species. Humans are sensitive to ITDs for tone frequencies up to about 1500 Hz. Neural sensitivity to ITD in cats, for example, goes up to 3kHz [50], and in owls, up to 6-8kHz [65].

Interaural Phase Difference (IPD)

Rose et al. [83] found that for some cells in the inferior colliculus of the cat the discharge rate was a periodic function of interaural time delay, with a period equal to the stimulus wavelength. They argued that the periodic nature of the response showed that the cells were sensitive to the interaural phase difference (IPD) rather than ITD. Similar cells were found in other parts of the mammalian auditory system (e.g., [30]). Of course, the issue of distinction between IPD and ITD arises only when the sound carrier frequency is changed, because, for example, if the cell is sensitive to ITD, then its preferred IPD (the one that elicits the largest response) would change linearly with carrier frequency; if, on the other hand, the cell is sensitive to IPD, its preferred IPD would stay the same. For pure tones of a fixed frequency, talking about IPD and ITD is redundant and we will sometimes use them interchangably, assuming that the stimulus is a pure tone at the characteristic frequency of the neuron.

Schematic of the Experiments

In experimental studies concerning the role of binaural cues and corresponding neuronal sensitivity, the sound is usually played from speakers located at one of the positions around the listener (free-field), or via headphones (dichotic). Headphone stimulation is, of course, an idealized sound representation (for example, the stimuli in the headphones are perceived as located within the head, [118]). On the other hand, it may be hard to control the parameters of the stimuli in the free-field studies. For example, the simplest geometrical model of IPD induction, due to Woodworth [110], that assumes constant speed and circular head shape, does not match the direct measurements of IPD [48] (Gourevitch [31] notes that for a 400 Hz signal at 15°off midline, the observed IPD is about three times greater than IPD based on the Woodworth's model). A more detailed model by Kuhn [49] takes into account physics of sound wave encountering the head as an obstacle. It shows the dependence of IPD on the head size and the sound carrier frequency for humans and provides better fit for the data, but also demonstrates that extracting the information available to the auditory system in a free-field study is not a trivial task.

Neuronal Mechanisms of IPD-Sensitivity: Superior Olive

Many contemporary models of binaural interaction are based on the idea due to Jeffress [38]. He suggested the following model: an array of neurons receives signals from the two ears through delay lines (axons that run along the array and provide conduction delay). The key assumption is that each neuron fires maximally when the signals from both ears arrive to that neuron simultaneously, in other words — when the conductance delay exactly compensates the IPD. These neurons are referred to as coincidence detectors. Mathematically, we can also think about their action as computing the temporal correlation of signals. The Jeffress-like arrangement was anatomically found in the avian homolog of medial superior olive (the first site of binaural interaction), the nucleus laminaris [117]. In mammals, recordings from single cells in MSO confirm that many cells do have the properties of coincidence detectors [30, 97, 112]. By recording the responses of these neurons to binaurally presented tones with different IPDs (simulating sounds at different locations around the head of the animal), one produces the (static) tuning curve (schematic in Fig. 16). It is usually a relatively smooth curve that has a clearly distinguishable peak. The peak of the binaural response appears when the ITD is approximately equal (with a different sign) to the difference between times of maximal responses to monaural signals [30], as expected for coincidence detectors.

In natural environments the relative positions of a sound source and the head are free to change and, therefore, the binaural sound localization cues are likely to be changing (dynamic) too. In experiments (e.g., [113]), the binaural cue produced by a moving sound source may be simulated by temporal variation of IPD. (In humans dynamic IPD modulation creates the percept of sound motion.) If the MSO cells act like coincidence detectors, they are expected to signal the instantaneous value of the interaural delay and show no evidence for motion sensitivity. This is confirmed experimentally (e.g., [98]).

Coincidence Detector Model

Cell Body

We describe here a coincidence detector point neuron model of the Morris-Lecar type (after Agmon-Snir et al. [2]). The main dynamic variables are the voltage V and a gating variable w.

$$C\dot{V} = -g_{Na}m_\infty(V)(V - V_{Na}) - g_K \cdot w \cdot (V - V_K) - g_L(V - V_L) - G_{syn}(V - V_{syn})$$

$$\tau_w(V)\dot{w} = w_\infty - w$$

Parameters are chosen in a realistic range to tune the model neuron into the phasic firing regime. Namely, this cell will fire only one spike in response to a depolarizing current injection (Fig. 17), as observed experimentally (e.g. [82]). Most of the parameters are the same as in Agmon-Snir et al. [2], unless noted otherwise.

Input

Input is modeled as 2 delay lines (one from each ear). Each line makes 12 synapses on the cell (Fig. 18A). On each stimulus cycle every synapse can produce a synaptic event with probability 0.7 (at 500 Hz stimulus frequency this corresponds to average firing rate of the afferents equal to 350 Hz) (Fig. 18B). In case a synaptic event is produced, it results in the alpha-function change of postsynaptic conductance, namely

$$G_{syn,unit} = A\frac{t - \phi}{t_p}\exp\left(1 - \frac{t - \phi}{t_p}\right).$$

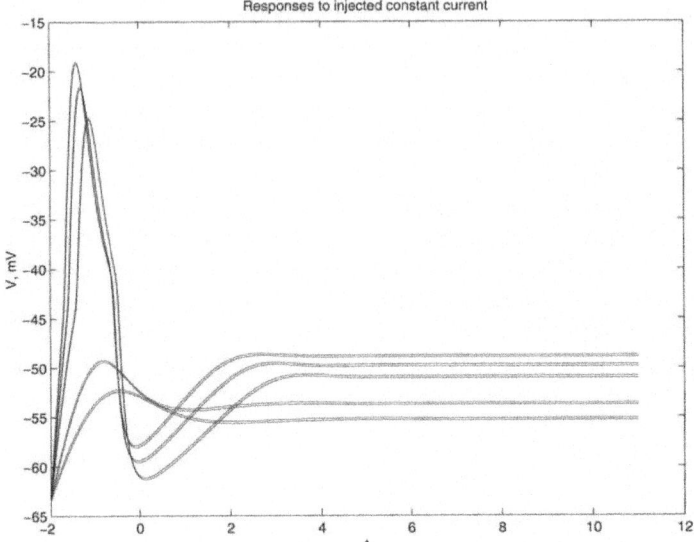

Fig. 17. Responses to increasing level of depolarizing current. The cell fires only once in response to a constant level injection of depolarizing superthreshold current.

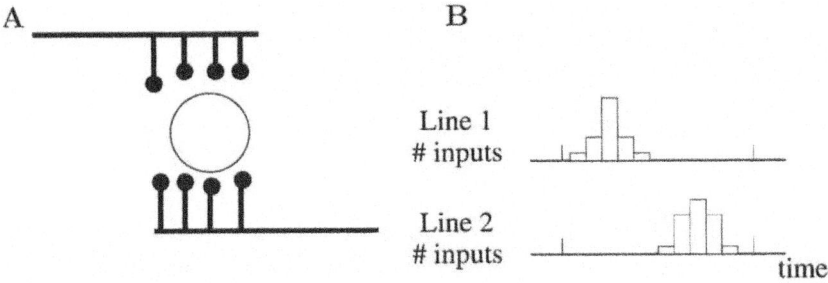

Fig. 18. Schematic of the input to the model cell. *A:* Delay lines. *B:* Example of distribution of input events times within stimulus cycle. Total number of events is also random, see text.

Here, $t_p = .1$ msec, amplitude A is chosen to produce reasonable firing rates at the given number of inputs and ϕ is chosen from a Gaussian distribution $N(\phi_0, \sigma^2)$ for one side and $N(\text{IPD}, \sigma^2)$ for the other side (IPD – interaural phase disparity, $\sigma = 0.4$ – measure of phase-locking, ϕ_0 – preferred phase) (see Fig. 19A).

Figures 19 and 20 show sample voltage and synaptic conductance time courses and a tuning curve of the model neuron. The tuning curve was obtained by computing number of spike elicited during a 1 sec stimulus presen-

A   **B**

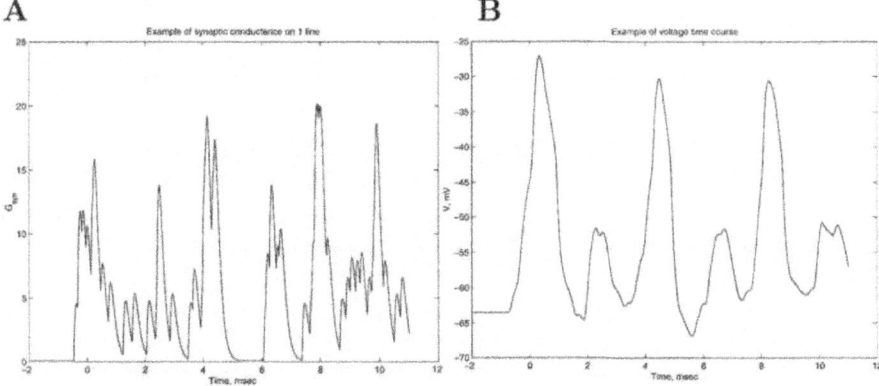

Fig. 19. Sample conductance and voltage traces. *A:* Synaptic conductance of one line of inputs over 6 stimulus cycles. *B:* Corresponding voltage trace.

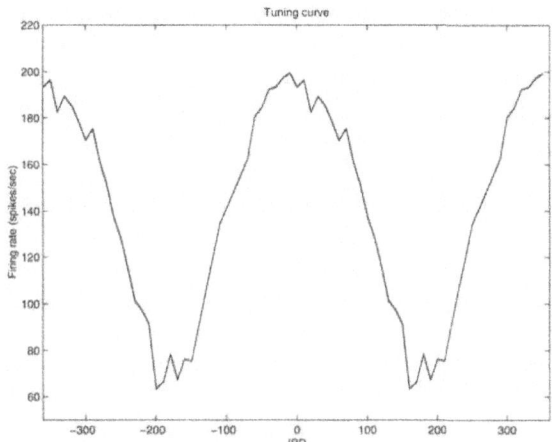

Fig. 20. Tuning curve of the model neuron. It is computed with small amount of averaging.

tation. This tuning curve shows that the neuron is a pretty good coincidence detector.

However, the known coincident-detector avian auditory neurons have a very specific structure. They have bipolar dendrites whose length varies continuously with characteristic frequency of the neuron (neurons with higher characteristic frequencies have shorter dendrites). The function of the dendrites was hard to unravel experimentally, until a mathematical model [2] suggested that propagation of signal along the dendrites resulted in nonlinear summation of conductances at the soma, improving coincidence detection. The argument goes as follows. Let us measure the quality of coincidence detection as the difference between maximum and minimum of the tuning curve. Let us also assume that the unitary synaptic conductance is adjusted in such a way that when signals from both sides arrive simultaneously the probability of firing is the same for either a point neuron or a bipolar neuron (this ensures that the maximum of the tuning curve is the same), i.e. let us say that the cell needs n simultaneous input spikes to fire ($n/2$ from each side). First, we need to show that the quality of coincidence detection is better for a neuron with bipolar dendrite than for a point neuron (i.e. minimum of the tuning curve is lower with the dendrite included). The key observation is that the synaptic current produced from a number of input spikes arriving at the same location is smaller than the current produced by the same number of inputs arriving at several different locations, because the driving force for excitatory current decreases with increase in local voltage. When the signals from different sides arrive half-cycle apart (worst ITD) and by chance one side produces n input spikes, it is still enough to fire a point neuron (which does not care where the inputs came from), but in the bipolar dendrite neuron all n spikes will now be concentrated on one side, producing smaller (insufficient) synaptic current, i.e. probability of firing will be smaller with dendrite included. Second, we need to show that longer dendrites improve coincidence detection at lower rather than higher input frequencies. This happens because of the jitter present in the input trains. Higher frequencies mean larger overlap between inputs from both sides even when the phase shift is maximal. These occasional "false" coincidences, will only add a small fraction to the conductance. For the point neuron it means only a small change in synaptic current. However, in case with dendrite the small conductance arrives at the different dendrite, producing relatively large current, i.e. higher probability of the "false" response. As a result, at a given level of jitter, the minimum of the tuning curve increases with frequency, i.e. the dendrites become less advantageous at higher frequencies.

In addition, more recent computational models of MSO argue that Jeffress model may need to be seriously revised. In fact, it has been shown experimentally that MSO cells that ate ITD-sensitive may be receiving substantial location-tuned inhibitory input [13]. A computational model [13] confirms that if that inhibition arrives at the MSO cell with a fixed time delay relative to excitation from the same side of the brain, it may allow to explain a larger range of MSO response properties than traditional excitation-only coincidence detector Jeffress-type model.

6 Midbrain

The inferior colliculus (IC) is the midbrain target of all ascending auditory information. It has two major divisions, the central nucleus and dorsal cortex, and both are tonotopically organized. The inputs from brainstem auditory nuclei create multiple tonotopic maps to form what are believed to be locally segregated functional zones for processing of different aspects of the auditory stimuli. The central nucleus receives both direct monaural input from cochlear nuclei and indirect binaural input from the superior olive.

6.1 Cellular Organization and Physiology of Mammalian IC

The IC appears to be an auditory integrative station as well as a switchboard. It is responsive to interaural delay and amplitude differences and may provide a spatial map of the auditory environment, although this has not been directly shown, except in some birds, e.g., barn owls. Changes in the spectrum of sounds such as amplitude and frequency modulation appear to have a separate representation in the IC. This sensitivity to spectral changes possibly provides the building blocks for neurons responsive to specific phonemes and intonations necessary to recognize speech. Finally, the IC is involved in integration and routing of multi-modal sensory perception. In particular, it sends projections which are involved in ocular reflexes and coordination between auditory and visual systems. It also modifies activity in regions of the brain responsible for attention and learning.

One of the very interesting properties of the inferior colliculus responses is its sensitivity to interaural time (or phase) differences. As we explained above, ITD sensitivity first arises in the MSO, due to precise preservation of timing information (phase locking) and coincidence detection. Notably, preservation of timing information starts to deteriorate in the IC. In particular even its low frequency cells do not phase lock as well as cells at the previous levels. Yet, in spite of apparent lack of timing information, new properties of ITD sensitivity, such as non-linear responses to dynamic changes in ITD arise in the inferior colliculus. This has recently attracted attention of modelers to the IC [14, 15, 88, 11, 12].

6.2 Modeling of the IPD Sensitivity in the Inferior Colliculus

The inferior colliculus (IC), in the midbrain, is a crucial structure in the central auditory pathway (Fig. 1). Ascending projections from many auditory brain stem nuclei converge there. Nevertheless, its role in the processing of auditory information is not well understood. Recent studies have revealed that many neurons in the IC respond differently to auditory stimuli presented under static or dynamic conditions and the response also depends on the history of stimulation. These include responses to dynamically varying binaural cues: interaural phase [62, 95, 96] and interaural level differences [87], and monaural stimuli such as slowly modulated frequency [57]. At least some of the sensitivity to dynamic temporal features is shown to originate in IC or in the projections to IC, because, as we mentioned above, responses of cells in lower level structures (MSO and dorsal nuclei of lateral lemniscus (DNLL) for the low-frequency pathway, Fig. 1) follow the instantaneous value of the stimulus [98]. A history-dependence of the responses, similar to that in IC.

but of even greater magnitude, has been observed in the auditory cortex [58]. These findings suggest that there might be a hierarchy of representations of auditory signals at subsequent levels of the auditory system.

Sensitivity to simulated motion in IC has been demonstrated with binaural beats (periodic change of IPD through 360°), [95, 114] and partial range sweeps (periodic modulation of interaural phase in a limited range) [96, 62]. For example, it was shown that the binaural beat response is sometimes shifted with respect to the static tuning curve (Fig. 22, 23), responses to IPD sweeps form hysteresis loops (Fig. 24), and the response to the dynamic stimulus can be strong outside of the statically determined excitatory receptive field (example in Fig. 24A, arrow). Thus, there is a transformation of response properties between MSO and IC, but there is no agreement on how this transformation is achieved. Both intrinsic and synaptic mechanisms have been implicated.

One hypothesis is that the observed history-dependence of responses is reflective of "adaptation of excitation" (firing rate adaptation) [62, 98]. The firing rate adaptation is a slow negative feedback mechanism that can have various underlying cellular or network mechanisms. It can manifest itself in extracellular recordings of responses to a constant stimulus as the decrease of probability of firing with time, and in *in vitro* recordings as the increase of the interspike interval in response to the injection of a constant depolarizing current. Both extracellular [62, 98], and intracellular [72, 93] recordings have registered firing rate adaptation in IC. It has also been shown by Cai et al., in a modeling study [15], that addition of a negative feedback, in the form of a calcium-activated hyperpolarizing membrane conductance (one of the adaptation mechanisms), to a motion-insensitive model of an IC neuron, induced sensitivity to dynamic stimuli.

Another hypothesis [98] is that the motion is the result of interaction between excitatory and inhibitory inputs, aided by adaptation and post-inhibitory rebound. Post-inhibitory rebound (PIR) is observed when the neuron fires upon its release from hyperpolarization; this has been demonstrated in IC slice recordings [72, 93]. McAlpine and Palmer [64] argued that leaving the key role in generation of motion sensitivity to inhibition contradicts their data. They showed that sensitivity to the apparent motion cues is decreased in the presence of the inhibitory transmitter, GABA, and increased in the presence of inhibitory transmission blocker, bicuculline.

Unification of the hypotheses. To examine the role of various mechanisms in shaping IC response properties, mathematical models of an IC cell were developed in which the cell receives IPD-tuned excitatory and inhibitory inputs and possesses the properties of adaptation and post-inhibitory rebound.

In earlier modeling studies of IC, Cai et al. [14, 15] developed detailed cell-based spike-generating models, involving multiple stages of auditory processing. One of their goals was to describe some of the data that require a hierarchy of binaural neurons. Recent models [11, 12] take a very different approach. These models are minimal in that they only involve the components whose role we want to test and do not explicitly include spikes (firing-rate-type models). Exclusion of spikes from the model is based on the following consideration. First, IC cells, unlike neurons in MSO, do not phase lock to the carrier frequency of the stimulus. In other words, spike timing in IC is less precise. Second, it is assumed that the convergent afferents from MSO represent a distribution of preferred phases and phase-locking properties. As a result, either through filtering by the IC cell or dispersion among the inputs, spike-timing information is not of critical importance to these IC cells. This minimalistic approach greatly facilitates the examination of computations that can be performed with rate-coded inputs (spike-timing precision on the order of 10 ms), consistent with the type of input information available to neurons at higher levels of processing. In addition, these models are computationally efficient, can be easily implemented, and require minimal assumptions about the underlying neural system.

It is shown in the modeling studies [11, 12] that intrinsic cellular mechanisms, such as adaptation and post-inhibitory rebound, together with the interaction of excitatory and inhibitory inputs, contribute to the shaping of dynamic responses. It is also shown that each of these mechanisms has a specific influence on response features. Therefore, if any of them is left out of consideration, the remaining two cannot adequately account for all experimental data.

Model

Cellular Properties

The model cell [12] represents a low-frequency-tuned IC neuron that responds to interaural phase cues. Since phase-locking to the carrier frequency among such cells is infrequent and not tight [51], this model assumes that binaural input is rate-encoded. In this model the spikes are eliminated by implicitly averaging over a short time scale (say, 10 ms). Here, we take as the cell's observable state variable its averaged voltage relative to rest (V). This formulation is in the spirit of early rate models (e.g., [32]), but extended to include intrinsic cellular mechanisms such as adaptation and rebound. Parameter values are not optimized, most of them are chosen to have representative values within physiological constraints, as specified in this section.

The current-balance equation for the IC cell is:

$$C\dot{V} = -g_L(V - V_L) - \bar{g}_a a(V - V_a) - I_{syn}(t) - I_{PIR} + I. \qquad (16)$$

The terms on the right-hand side of the equation represent (in this order) the leakage, adaptation, synaptic, post inhibitory rebound, and applied currents. The applied current I is zero for the examples considered here and all other terms are defined below.

Leakage current has reversal potential $V_L = 0$ mV and conductance $g_L = 0.2$ mS/cm^2. Thus, with $C = 1$ μF/cm^2, the membrane time constant $\tau_V = C/g_L$ is 5 ms.

The adaptation is modeled by a slowly-activating voltage-gated potassium current $\bar{g}_a a (V - V_a)$. Its gating variable a satisfies

$$\tau_a \dot{a} = a_\infty(V) - a. \tag{17}$$

Here $a_\infty(V) = 1/(1 + \exp(-(V - \theta_a)/k_a))$, time constant $\tau_a = 150$ ms, $k_a = 5$ mV, $\theta_a = 30$ mV, maximal conductance $\bar{g}_a = 0.4$ mS/cm^2, reversal potential $V_a = -30$ mV. The parameters k_a, θ_a are chosen so that little adaptation occurs below firing threshold; and $\bar{g}_a$ so that when fully adapted for large inputs, the firing rate is reduced by 67%. The value of τ_a, assumed to be voltage-independent, matches the adaptation time scale range seen in spike trains (Semple, unpublished) and it leads to dynamic effects that provide good comparison with results over the range of stimulus rates used in experiments.

The model's readout variable, r, represents firing rate, normalized by an arbitrary constant. It is an instantaneous threshold-linear function of V: $r = 0$ for $V < V_{th}$ and $r = K \cdot (V - V_{th})$ for $V \geq V_{th}$, where $V_{th} = 10$ mV. We set (as in [12]) $K = 0.04075$ mV^{-1}. Its value does not influence the cell's responsiveness, only setting the scale for r.

Synaptic Inputs

As proposed by Spitzer and Semple [98] and implemented in the spike-based model of Cai et al. [14, 15], this model IC cell receives binaural excitatory input from neurons in ipsilateral MSO (e.g., [1, 70]) and indirect (via DNLL [91]) binaural inhibitory input from the contralateral MSO. Thus DNLL, although it is not specifically included in the model, is assumed (as in Cai et al. [14]) to serve as an instantaneous converter of excitation to inhibition.

The afferent input from MSO is assumed to be tonotopically organized and IPD-tuned. Typical tuning curves of MSO neurons are approximately sinusoidal with preferred (maximum) phases in contralateral space [97, 112]. We use a sign convention that the IPD is positive if the phase is leading in the ear contralateral to the IC cell, to choose the preferred phases $\phi_E = 40°$ for excitation and $\phi_I = -100°$ for inhibition. Preferred phases of MSO cells are broadly distributed [63, 58]. Similar ranges work in the model, and in that sense, the particular values picked are arbitrary.

While individual MSO cells are phase-locked to the carrier frequency [97, 112], we assume that the effective input to an IC neuron is rate-encoded, as explained above. Under this assumption, the synaptic conductance transients from incoming MSO spikes are smeared into a smooth time course that traces the tuning curve of the respective presynaptic population (Fig. 21). Therefore, the smoothed synaptic conductances, averaged over input lines and short time scales, g_E and g_I, are proportional to the firing rates of the respective MSO populations and they depend on t if IPD is dynamic. Then

$$I_{syn} = -g_E(V - V_E) - g_I(V - V_I), \tag{18}$$

$$g_{E,I} = \bar{g}_{E,I} \left[0.55 + 0.45 \cos(IPD(t) - \phi_{E,I}) \right]. \tag{19}$$

Reversal potentials are $V_E = 100$ mV, $V_I = -30$ mV, maximum conductance values $\bar{g}_E = 0.3$ mS/cm^2, $\bar{g}_I = 0.43$ mS/cm^2.

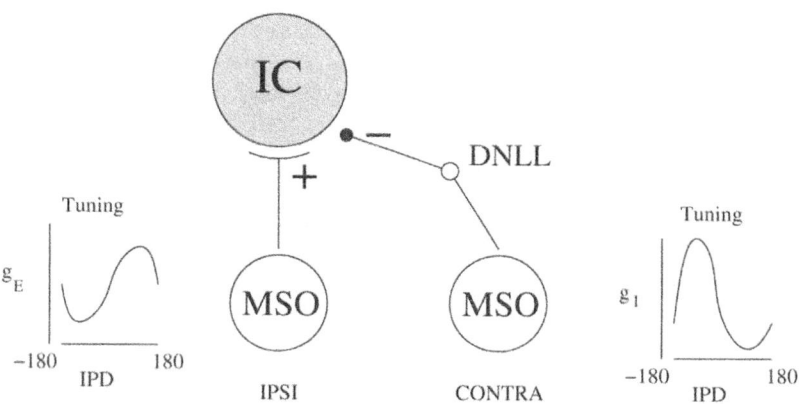

Fig. 21. Schematic of the model. A neuron in the inferior colliculus (marked IC), receiving excitatory input from ipsilateral medial superior olive (MSO); contralateral input is inhibitory. Each of the MSO-labelled units represents an average over several MSO cells with similar response properties (see text). DNLL (dorsal nucleus of lateral lemniscus) is a relay nucleus, not included in the model. Interaural phase difference (IPD) is the parameter whose value at any time defines the stimulus. The tuning curves show the stimulus (IPD) tuning of the MSO populations. They also represent dependence of synaptic conductances (g_E and g_I) on IPD (see text). Used by permission of The American Physiological Society from [12].

Rebound Mechanism

The post-inhibitory rebound (PIR) mechanism is implemented as a transient inward current (I_{PIR} in the equation (16)):

$$I_{PIR} = g_{PIR} \cdot m \cdot h \cdot (V - V_{PIR}). \tag{20}$$

The current's gating variables, fast (instantaneous) activation, m, and slow inactivation, h, satisfy:

$$m = m_\infty(V), \tag{21}$$
$$\tau_h \dot{h} = h_\infty(V) - h, \tag{22}$$

where $m_\infty(V) = 1/(1 + e^{-(V-\theta_m)/k_m})$, $h_\infty(V) = 1 - 1/(1 + e^{-(V-\theta_h)/k_h})$, $\tau_h = 150$ ms and $V_{PIR} = 100$ mV. The parameter values: $k_m = 4.55$ mV, $\theta_m = 9$ mV, $k_h = .11$ mV, $\theta_h = -11$ mV, are chosen to provide only small current at steady state for any constant V and maximum responsiveness for voltages near rest. Maximum conductance g_{PIR} equals 0.35 mS/cm^2 in Fig. 24 and zero elsewhere.

Stimuli

The only sound input parameter that is varied is IPD as represented in equations (18) and (19); the sound's carrier frequency and pressure level are fixed. The static tuning curve is generated by presentation of constant-IPD stimuli. The dynamic stimuli are chosen from two classes. First, the binaural beat stimulus is generated as: $IPD(t) = 360° \cdot f_b \cdot t$ (mod 360°), with beat frequency f_b that can be negative or positive. Second, the partial range sweep is generated by $IPD(t) = P_c + P_d \cdot \text{triang}(P_r t/360)$. Here, triang($\cdot$) is a periodic function (period 1) defined by triang(x) = $4x - 1$ for $0 \leq x < 1/2$ and triang(x) = $-4x + 3$ for $1/2 \leq x < 1$. The stimulus parameters are the sweep's center P_c; sweep depth P_d (usually 45°); and sweep rate P_r (in degrees per second, usually 360°/s). We call the sweep's half cycle where IPD increases – the "up-sweep", and where IPD decreases – the "down-sweep".

Simulation Data Analysis

The response's relative phase (Figs. 22 and 23) is the mean phase of the response to a binaural beat stimulus minus the mean phase of the static tuning curve. The mean phase is the direction of the vector that determines the response's vector strength [30]. To compute the mean phase we collect a number of response values (r_j) and corresponding stimulus values (IPD_j in radians). The average phase φ is such that $\tan \varphi = (\sum r_j \sin(IPD_j))/(\sum r_j \cos(IPD_j))$. For the static tuning curve the responses were collected at IPDs ranging from -180° to 180° in 10° intervals. For the binaural beat the response to the last stimulus cycle is used.

Example of Results: Binaural Beats

The binaural beat stimulus has been used widely in experimental studies (e.g., [52, 61, 96, 107] to create interaural phase modulation: a linear sweep through a full range of phases (Fig. 22). Positive beat frequency means IPD increases during each cycle; negative beat frequency — IPD decreases (Fig. 22A). Using the linear form of IPD(t) to represent a binaural beat stimulus in the model, we obtain responses whose time courses are shown in Fig. 22A, lower.

To compare responses between different beat stimuli and also with the static response, we plot the instantaneous firing rate vs. the corresponding IPD. Fig. 22B shows an example that is typical of many experimentally recorded responses (e.g., [52, 95, 96]). Replotting our computed response from 27A in 27C reveals features in accord with the experimental data (compare 27B and 27C): dynamic IPD modulation increases the sharpness of tuning; preferred phases of responses to the two directions of the beat are shifted in opposite directions from the static tuning curve; maximum dynamic responses are significantly higher than maximum static responses. These features, especially the phase shift, depend on beat frequency (Fig. 22C, D).

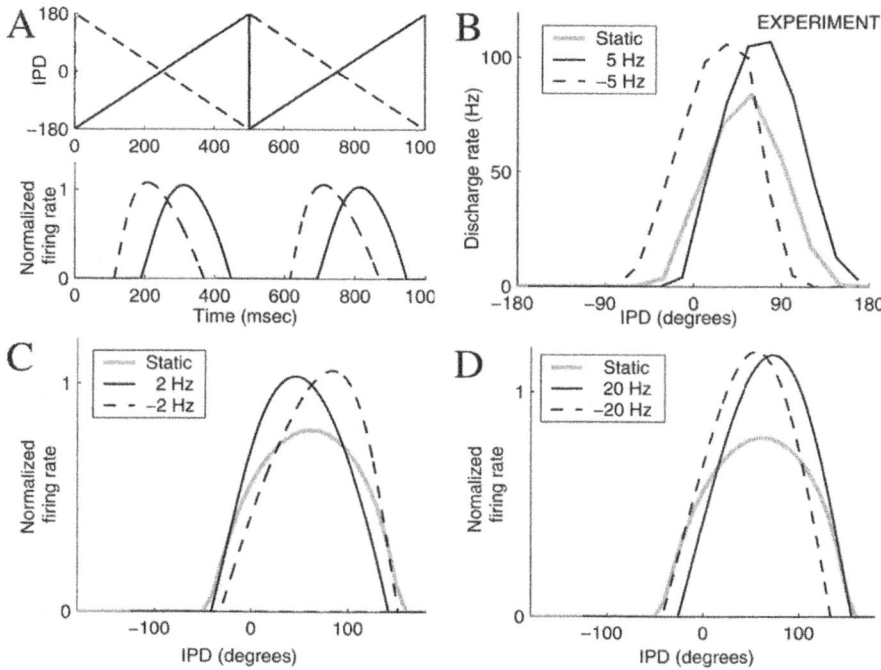

Fig. 22. Responses to binaural beat stimuli. *A:* Time courses of IPD stimuli (upper) and corresponding model IC responses (lower). Beat frequencies 2 Hz (solid) and -2 Hz (dashed). *B-D:* Black lines show responses to binaural beats of positive (solid) and negative (dashed) frequencies. Grey line is the static tuning curve. *B:* Example of experimentally recorded response. Carrier frequency 400 Hz, presented at binaural SPL 70 dB. Beat frequency 5 Hz. *C,D:* Model responses over the final cycle of stimulation vs corresponding IPD. Beat frequency is equal to 2 and 20 Hz, respectively. Used by permission of The American Physiological Society from [12].

Dependence of Phase on the Beat Frequency

In Fig. 22D the direction of phase shift is what one would expect from a lag in response. Fig. 22C is what would be expected in the presence of adaptation (e.g., [94]; as the firing rate rises, the response is "chopped off" by adaptation at the later part of the cycle. This shifts the average of the response to earlier phases). In particular, if we interpret the IPD change as an acoustic motion, then for a stimulus moving in a given direction there can be a phase advance or phase-lag, depending on the speed of the simulated motion (beat frequency).

To characterize the dependence of the phase shift on parameters, we define the response phase as in a vector-strength computation (see above). We take the average phase of the static response as a reference ($\varphi_0 = 0.1623$). Fig. 23A shows the relative phase of the response ($\tilde{\varphi}_{f_b} - \varphi_0$) vs. beat frequency ($f_b$). At the smallest beat frequencies the tuning is close to the static case and the relative phase is close to zero. As the absolute value of beat frequency increases, so does the phase advance. The phase-advance is due to the firing rate adaptation (see above). At yet higher frequencies a phase lag develops. This reflects the RC-properties of the cell membrane — the membrane time constant prevents the cell from responding fast enough to follow the high frequency phase modulation.

Plots such as in Fig. 23A have been published for several different IC neurons [98, 12] and examples are reproduces in Fig. 23B. Whereas the present model does not have a goal of quantitative match between the modeling results and experimental data, there are two qualitative differences between Figs. 23A and 23B. First, the range of values of the phase shift is considerably larger in the experimental recordings, particularly, the phase lag at higher beat frequencies. Second, it had been reported in the experimental data, that there is no appreciable phase advance at lower beat frequencies. Fig. 23B suggests that the phase remains approximately constant before shifting in the positive direction. Only in the close-up view of the phase at low beat frequencies (Fig. 23B, inset) can the phase-advance be detected. It is a surprising observation, given that the firing rate, or spike frequency, adaptation has been reported in IC cells on many occasions (e.g., [62, 96]) and that, as we noted above, the phase advance is a generic feature of cells with adapting spike trains.

Role of Transmission Delay

The modeling study [12] suggested that the absence of phase advance and high values of phase lag should be attributed to the transmission delay (five to tens of milliseconds) that exists in transmission of the stimulus to the IC. This delay was not included in the model, as outlined in *Model* above. In fact, we can find its contribution analytically (Fig. 23C, inset). Consider the response without transmission delay to the beat of positive frequency f_b . In the phase computation, the response at each of the chosen times (each bin of PSTH) is considered as a vector from the origin with length proportional to the recorded firing rate and pointing at the corresponding value of IPD. In polar coordinates $\tilde{\mathbf{r}}_j = (r_j, IPD_j)$, where r_j is the jth recorded data point (normalized firing rate), IPD_j is the corresponding IPD. The average phase of response ($\tilde{\varphi}_{f_b}$) can be found as a phase of the sum of these vectors. If the response is delayed, then the vectors should be $\mathbf{r}_j = (r_j, IPD_j + d \cdot f_b \cdot 360°)$, because the same firing rates would correspond to different IPDs (see Fig. 23C, inset). The vectors are the same as without transmission delay, just rotated by $d \cdot f_b$ fractions of the cycle. Therefore, the phase with transmission delay $\varphi_{f_b} = \tilde{\varphi}_{f_b} + d \cdot f_b$. If we plot the relative phase with the transmission delay, it will be $\varphi_{f_b} - \varphi_0 = (\tilde{\varphi}_{f_b} - \varphi_0) + d \cdot f_b$ The phase of the static tuning curve '(φ_0) does not depend on the transmission delay. Fig. 23C shows the graph from 23A right, modified by various transmission delays (thick curve is without delay — same as in 23A right). With transmission delay included in the model, the range of observed phase shifts increases dramatically and

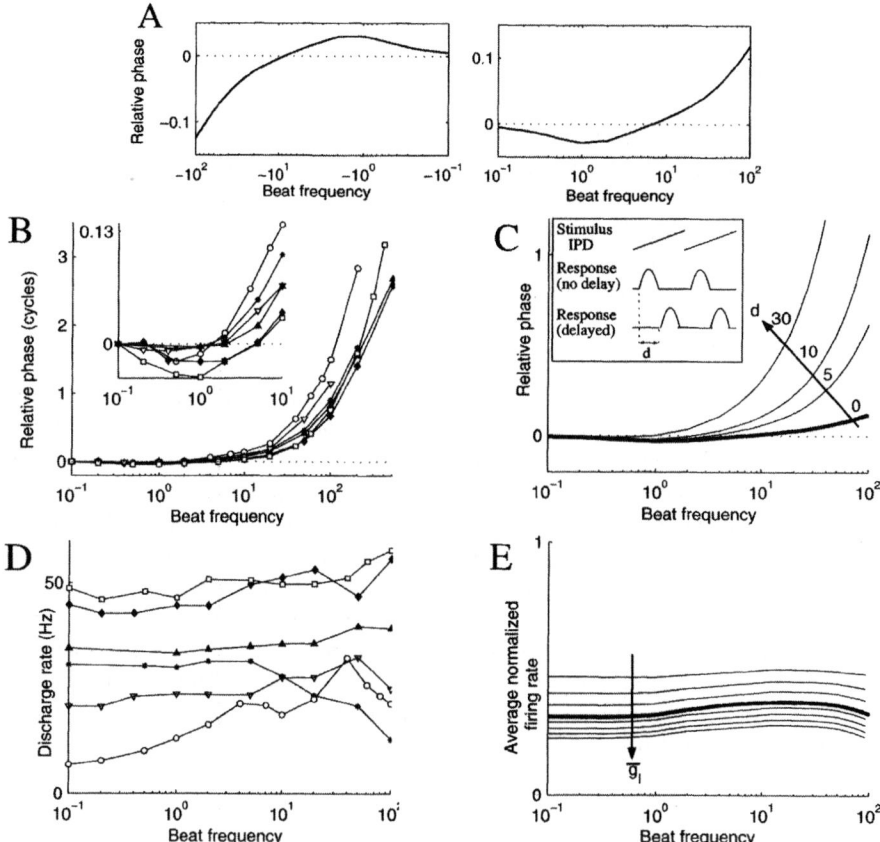

Fig. 23. *A:* Dependence of phase shift on beat frequency in the model neuron. Phase shift is relative to the static tuning curve (see text). *B:* Experimentally recorded phase (relative to the phase at 0.1 Hz beat frequency) vs beat frequency. Six different cells are shown. Carrier frequencies and binaural SPLs are: 1400 Hz, 80 dB (star); 1250 Hz, 80 dB (circle); 200 Hz, 70 dB (filled triangle); 800 Hz, 80 dB (diamond); 900 Hz, 70 dB (empty triangle); 1000 Hz, 80 dB (square). Inset is a zoom-in at low beat frequencies. *C:* Phase of the model response, recomputed after accounting for transmission delay from outer ear to IC, for various values of delay (*d*), as marked. Inset illustrates computation of delay-adjusted phase. *D,E:* Average firing rate as a function of beat frequency: experimental (*D*, same cells with the same markers as in B) and in the model (*E*). Multiple curves in *E* – for different values of $\bar{g}_I$ ($\bar{g}_I = 0.1$, 0.2, 0.3, 0.43, 0.5, 0.6, 0.7, 0.8 mS/cm^2). Thick line — for the parameter values given in *Methods*. Used by permission of The American Physiological Society from [12].

the phase-advance is masked. In practice neurons are expected to have various transmission delays, which would result in a variety of phase-frequency curves, even if the cellular parameters were similar.

Mean Firing Rate

Another interesting property of response to binaural beats is that the mean firing rate stays approximately constant over a wide range of beat frequencies (Fig. 23D and [98]). The particular value, however, is different for different cells. The same is valid for the model system (Fig. 23E). The thick curve is the firing rate for the standard parameter values. It is nearly constant across all tested beat frequencies (0.1 to 100 Hz). The constant value can be modified by change in other parameters of the system, e.g. $\bar{g}_I$ (shown).

Other Results

The model that we have just described allows to show [12] that the presence of firing rate adaptation together with IPD-tuned excitatory and inhibitory inputs can explain sharper tuning and phase shifts in response to beats; dis-contiguous responses to overlapping IPD sweeps; and a decrease in motion sensitivity in the presence of added inhibition. Also, a post-inhibitory re-bound mechanism (modeled as a transient membrane current activated by release from prolonged hyperpolarization) allows to explain the strong exci-tatory dynamic response to sweeps in the silent portion of the static tuning curve (Fig. 24A,B arrows). It also allows to make predictions for future *in vitro* and *in vivo* experiments and suggest experiments that might help to clarify what are the sources of inhibition. For further details and discussion of this model see [12].

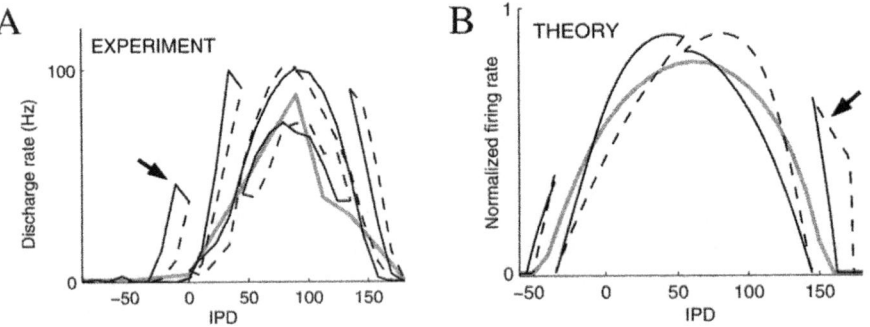

Fig. 24. Responses to partial range sweeps. "Rise-from-nowhere". *A:* Experimentally observed response arising in silent region of the static tuning curve (marked with an arrow). Carrier frequency 1050 Hz; binaural SPL 40 Hz. *B:* Model responses to various sweeps. The rightmost sweep produces "rise-from-nowhere" (arrow). Used by permission of The American Physiological Society from [12].

7 Thalamus and Cortex

The auditory part of the thalamus is the medial geniculate body. It is situated in the caudal part of the thalamus and is the last major relay station for ascending auditory fibers before they reach the cortex. There is a tonotopic arrangement in the medial geniculate body in which low frequencies are represented laterally and high frequencies are located medially in the principal part. The main projection of the medial geniculate body is to the primary auditory cortex. The medial geniculate body also sends fibers to other thalamic nuclei and may play a part in a regulatory feedback system, with descending projections to the inferior colliculus, the nucleus of the lateral lemniscus, the trapezoid body, and the superior olivary nucleus.

The primary auditory cortex is a primary auditory reception area in cortex and it receives its projections from the medial geniculate body. Areas immediately adjacent to the primary auditory cortex are auditory-association areas. These association areas receive signals from the primary auditory cortex and send projections to other parts of the cortex. Tonotopic organization of the auditory cortex is particularly complex. In the simplest analysis, high frequencies are represented anteriorly and low frequencies posteriorly in the auditory cortex. Each auditory cortical area is reciprocally connected to an area of the same type area in the contralateral hemisphere. In addition, auditory-association areas connect with other sensory-association areas such as somatosensory and visual. They also send projections that converge in the parietotemporal language area. It appears that the higher level of integration in the association areas is responsible for more complex interpretation of sounds. These properties of the auditory cortex may explain why patients with lesions in one of the cortical hemispheres have little difficulty with hearing as measured by presenting simple sounds, e.g., pure tones. However, such patients may have impaired ability to discriminate the distorted or interrupted speech patterns and have difficulty focusing on one stimulus if a competing message, especially meaningful, i.e., speech is presented at the same time.

There is also a descending efferent auditory pathway that parallels the afferent pathway and is influenced by ascending fibers via multiple feedback loops. The specific function of this system in audition is not well understood, but clearly modulates central processing and regulates the input from peripheral receptors.

Because auditory cortex seems to deal primarily with higher brain functions, its modeling has been very scarce. However, recently there has been an promising set of studies by Middlebrooks and colleagues, in which they teach neural networks to interpret cortical recordings and make predictions of the psychophysical performance [24]. More details will have to be unravelled about biophysical properties of cortical auditory cells before more detailed modeling can be successfully used.

References

1. Adams JC. Ascending projections to the inferior colliculus, J. Comp. Neurol. 183: 519-538, 1979

2. Agmon-Snir H, Carr CE, Rinzel J. A case study for dendritic function: improving the performance of auditory coincidence detectors, Nature 393: 268-272, 1998

3. Allen JB. Two-dimensional cochlear fluid model: new results, J. Acoust. Soc. Am. 61: 110-119, 1977

4. Allen JB, Sondhi MM. Cochlear macromechanics: time domain solutions, J. Acoust. Soc. Am. 66: 123-132, 1979

5. Anderson DJ, Rose JE, Hind JE, Brugge JF. Temporal position of discharges in single auditory nerve fibers within the cycle of a sine-wave stimulus: frequency and intensity effects, J. Acoust. Soc. Am. 49: 1131-1139, 1971

6. Banks MI and Sachs MB. Regularity analysis in a compartmental model of chopper units in the anteroventral cochlear nucleus. J. Neurophysiol. 65: 606-629, 1991

7. Beyer RP. A computational model of the cochlea using the immersed boundary method, J. Comp. Phys. 98: 145-162, 1992

8. Blackburn CC and Sachs MB. Classification of unit types in the anteroventral cochlear nucleus: PST histograms and regularity analysis, J. Neurophys. 62: 1303-1329, 1989

9. Blum JJ, Reed MC, Davies JM. A computational model for signal processing by the dorsal cochlear nucleus. II. Responses to broadband and notch noise. J. Acoust. Soc. Am. 98: 181-91, 1995

10. Bogert BP. Determination of the effects of dissipation in the cochlear partition by means of a network representing the basilar membrane, J. Acoust. Soc. Am. 23: 151-154, 1951

11. Borisyuk A, Semple MN, Rinzel J. Computational model for the dynamic aspects of sound processing in the auditory midbrain, Neurocomputing 38: 1127-1134, 2001

12. Borisyuk A, Semple MN and Rinzel J. Adaptation and Inhibition Underlie Responses to Time-Varying Interaural Phase Cues in a Model of Inferior Colliculus Neurons, J. Neurophysiol. 88: 2134-2146, 2002

13. Brand A, Behrend O, Marquardt T, McAlpine D, Grothe B. Precise inhibition is essential for microsecond interaural time difference coding, Nature 417: 543-547, 2002

14. Cai H, Carney LH, Colburn HS. A model for binaural response properties of inferior colliculus neurons. I. A model with interaural time difference-sensitive excitatory and inhibitory inputs. J. Acoust. Soc. Am. 103: 475-493, 1998a

15. Cai H, Carney LH, Colburn HS. A model for binaural response properties of inferior colliculus neurons. II. A model with interaural time difference-sensitive excitatory and inhibitory inputs and an adaptation mechanism. J. Acoust. Soc. Am. 103: 494-506, 1998b

16. Carney LH. A model for the responses of low-frequency auditory-nerve fibers in cat. J. Acoust. Soc. Am. 93: 401-417, 1993

17. Chadwick RF. Studies in cochlear mechanics, in: M.H. Holmes, A. Rubenfeld (Eds.), Mathematical Modeling of the Hearing Process, Lecture Notes in Biomathematics, vol. 43, Springer, Berlin, 1980.

18. Daniel SJ, Funnell WR, Zeitouni AG, Schloss MD, Rappaport J. Clinical applications of a finite-element model of the human middle ear. J. Otolaryngol. 30: 340-346, 2001
19. de Boer BP. Solving cochlear mechanics problems with higher order differential equations. J. Acoust. Soc. Am. 72: 1427-1434, 1982
20. Digital Anatomist Project, Department of Biological Structure, University of Washington, http://www9.biostr.washington.edu
21. Feddersen WE, Sandal ET, Teas DC, Jeffress LA. Localization of high frequency tones. J. Acoust. Soc. Am. 29: 988-991, 1957
22. Ferris P, Prendergast PJ. Middle-ear dynamics before and after ossicular replacement. J. Biomech. 33: 581-590, 2000
23. Fletcher H. On the dynamics of the cochlea. J. Acoust. Soc. Am. 23: 637-645, 1951
24. Furukawa S, Xu L, and Middlebrooks JC. Coding of Sound-Source Location by Ensembles of Cortical Neurons. J. Neurosci. 20: 1216 - 1228, 2000
25. Gan RZ, Sun Q, Dyer RK Jr, Chang KH, Dormer KJ. Three-dimensional modeling of middle ear biomechanics and its applications. Otol. Neurotol. 23: 271-280, 2002
26. Geisler CD. From sound to synapse: physiology of the mammalian ear. Oxford University Press, 1998
27. Gil-Carcedo E, Perez Lopez B, Vallejo LA, Gil-Carcedo LM, Montoya F. Computerized 3-D model to study biomechanics of the middle ear using the finite element method. Acta Otorrinolaringol. Esp. 53:527-537, 2002. Spanish.
28. Givelberg E, Bunn J. A Comprehensive Three-Dimensional Model of the Cochlea. J. Comput. Phys. 191: 377-391, 2003
29. Godfrey DA, Kiang NYS, and Norris BE. Single unit activity in the dorsal cochlear nucleus of the cat. J. Comp. Neurol. 162: 269-284, 1975
30. Goldberg J and Brown P. Response of binaural neurons of dog superior olivary complex to dichotic tonal stimuli some physiological mechanisms for sound localization. J. Neurophysiol. 32: 613-636, 1969
31. Gourevitch G. Binaural hearing in land mammals. In: Yost WA and Gourevitch G (Eds.) Directional Hearing, Springer-Verlag, 1987
32. Grossberg S. Contour enhancement, short term memory, and constancies in reverberating neural networks. Stud Appl Math 52: 213-257, 1973
33. Harris DM. Action potential suppression, tuning curves and thresholds: comparison with single fiber data. Hear. Res. 1: 133-154, 1979
34. Heinz MG, Colburn HS, and Carney LH. Evaluating auditory performance limits: I. One-parameter discrimination using a computational model for the auditory nerve. Neural Computation 13: 2273-2316, 2001
35. Heinz MG, Colburn HS, and Carney LH. Evaluating auditory performance limits: II. One-parameter discrimination with random level variation. Neural Computation 13, 2317-2339, 2001
36. Holmes MN. A mathematical model of the dynamics of the inner ear, J. Fluid Mech. 116: 59-75, 1982
37. Inselberg A, Chadwick RF. Mathematical model of the cochlea. i: formulation and solution, SIAM J. Appl. Math. 30: 149-163, 1976
38. Jeffress LA. A place theory of sound localization. J. Comp. Physiol. Psych. 41: 35-39, 1948

39. Johnson, DH. The relationship between spike rate and synchrony in responses of auditory-nerve fibers to single tones. J. Acoust. Soc. Am. 68: 1115-1122, 1980

40. Keener JP, Sneyd J. Mathematical physiology, chapter 23. Springer-Verlag, 1998.

41. Kelly DJ, Prendergast PJ, Blayney AW. The effect of prosthesis design on vibration of the reconstructed ossicular chain: a comparative finite element analysis of four prostheses. Otol. Neurotol. 24: 11-19, 2003

42. Kiang NYS, Morest DK, Godfrey DA, Guinan JJ, and Kane EC. Stimulus coding at coudal levels of the cat's auditory nervous system: I. Response characteristics of single units. In Basic Mechanisms in Hearing (Moller AR and Boston P., eds.) New York Academic, pp. 455-478, 1973

43. Kim DO, Milnar CE, Pfeiffer RR. A system of nonlinear differential equations modeling basilar membrane motion. J. Acoust. Soc. Am. 54: 1517-1529, 1973

44. Koike T, Wada H, Kobayashi T. Modeling of the human middle ear using the finite-element method. J. Acoust. Soc. Am. 111: 1306-1317, 2002

45. Kolston PJ. Comparing in vitro, in situ and in vivo experimental data in a three dimensional model of mammalian cochlear mechanics, Proc. Natl. Acad. Sci. USA 96: 3676-3681, 1999

46. Konishi M. Study of sound localization by owls and its relevance to humans. Comp. Biochem. Phys. A 126: 459-469, 2000

47. Kuhlmann L, Burkitt AN, Paolini A, Clark GM. Summation of spatiotemporal input patterns in leaky integrate-and-fire neurons: application to neurons in the cochlear nucleus receiving converging auditory nerve fiber input. J. Comput. Neurosci. 12: 55-73, 2002

48. Kuhn GF. Model for the interaural time differences in the azimuthal plane. J. Acoust. Soc. Am. 62: 157-167, 1977

49. Kuhn GF. Physical acoustics nd measurments pertaining to directional hearing. In: Yost WA and Gourevitch G (Eds.) Directional Hearing, Springer-Verlag, 1987

50. Kuwada S and Yin TCT. Physiological studies of directional hearing. In: Yost WA and Gourevitch G (Eds.) Directional Hearing, Springer-Verlag, 1987

51. Kuwada S, Yin TCT, Syka J, Buunen TJF, Wickesberg RE. Binaural interaction in low-frequency neurons in inferior colliculus of the cat. IV. Comparison of monaural and binaural response properties. J. Neurophysiol. 51: 1306-1325, 1984

52. Kuwada S, Yin TCT, Wickesberg RE. Response of cat inferior colliculus neurons to binaural beat stimuli: possible mechanisms for sound localization. Science 206 (4418): 586-588, 1979

53. Lesser MB, Berkley DA. Fluid mechanics of the cochlea. Part i, J. Fluid. Mech. 51: 497-512, 1972

54. Leveque RJ, Peskin CS, Lax PD. Solution of a two-dimensional cochlea model using transform techniques. SIAM J. Appl. Math. 45: 450-464, 1985

55. Leveque RJ, Peskin CS, Lax PD. Solution of a two-dimensional cochlea model with fluid viscosity. SIAM J. Appl. Math. 48: 191-213, 1988

56. Loh CH. Multiple scale analysis of the spirally coiled cochlea, J. Acoust. Soc. Am. 74: 95-103, 1983

57. Malone BJ and Semple MN. Effects of auditory stimulus context on the representation of frequency in the gerbil inferior colliculus. J. Neurophysiol. 86: 1113-1130, 2001

58. Malone BJ, Scott BH and Semple MN. Context-dependent adaptive coding of interaural phase disparity in the auditory cortex of awake macaques. J. Neurosci. 22, 2002

59. Manis PB and Marx SO. Outward currents in isolated ventral cochlear nucleus neurons, J. Neurosci. 11: 2865-2880, 1991

60. Manoussaki D, Chadwick RS. Effects of geometry on fluid loading in a coiled cochlea, SIAM J. Appl. Math. 61: 369-386, 2000

61. McAlpine D, Jiang D, Shackleton TM, Palmer AR. Convergent input from brainstem coincidence detectors onto delay-sensitive neurons in the inferior colliculus. J. Neurosci. 18: 6026-6039, 1998

62. McAlpine D, Jiang D, Shackleton TM, Palmer AR. Responses of neurons in the inferior colliculus to dynamic interaural phase cues: evidence for a mechanism of binaural adaptation. J. Neurophysiol. 83: 1356-1365, 2000

63. McAlpine D, Jiang D, Palmer AR. A neural code for low-frequency sound localization in mammals. Nat. Neurosci. 4: 396-401, 2001

64. McAlpine D and Palmer AR. Blocking GABAergic inhibition increases sensitivity to sound motion cues in the inferior colliculus. J. Neurosci. 22: 1443-1453, 2002

65. Moiseff A. and Konishi M. Neuronal and behavioral sensitivity to binaural time differences in the owl. J. Neurosci. 1: 40-48, 1981

66. Moller H, Sorensen MF, Hammershoi D and Jensen CB. Head-related transfer functions of human subjects. J. Audio Eng. Soc. 43: 300-321, 1995

67. Moore BCJ. An introduction to the psychology of hearing, 5th ed. Academic Press, 2003

68. Musicant AD, Chan JCK, and Hind JE. Direction-dependent spectral properties of cat external ear: new data and cross-species comparisons. J. Acoust. Soc. Am. 87: 757-781, 1990

69. Neu JC and Keller JB. Asymptotoic analysis of a viscous cochlear model. J. Acoust. Soc. Am. 1985

70. Oliver DL, Beckius GE, and Shneiderman A. Axonal projections from the lateral and medial superior olive to the inferior colliculus of the cat: a study using electron microscopic autoradiography. J. Comp. Neurol. 360: 17-32, 1995

71. Parham K and Kim DO. Spontaneous and sound-evoked discharge characteristics of complex-spiking neurons in the dorsal cochlear nucleus of the unanasthetized decerebrate cat. J. Neurophys. 73: 550-561, 1975

72. Peruzzi D, Sivaramakrishnan S, Oliver DL. Identification of cell types in brain slices of the inferior colliculus. J. Neurosci. 101: 403-416, 2000

73. Peskin CS. Partial differential equations in biology. Courant Inst. Lecture Notes, Courant Inst. of Mathematical Sciences, NYU, New York, 1975-76

74. Peskin CS. Lectures on mathematical aspects of physiology. Lectures in Appl. Math. 19: 38-69, 1981

75. Peskin CS. The immersed boundary method, Acta Numerica 11, 479-517, 2002

76. Peterson LC, Bogert BP. A dynamical theory of the cochlea, J. Acoust. Soc. Am. 22: 369-381, 1950

77. Popper AN, Fay RR, eds. The mammalian auditory pathway: neurophysiology. New York : Springer-Verlag, 1992

78. Ranke OF. Theory of operation of the cochlea: a contribution to the hydrodynamics of the cochlea. J. Acoust. Soc. Am. 22: 772-777, 1950

79. Rayleigh L. On our perception of sound direction. Phil. Mag. 13: 214-232, 1907

80. Reed MC, Blum JJ. A computational model for signal processing by the dorsal cochlear nucleus. I. Responses to pure tones. J. Acoust. Soc. Am. 97: 425-438, 1995

81. Reed MC, Blum J. A model for the computation and encoding of azimuthal information by the Lateral Superior Olive. J. Acoust. Soc. Amer. 88, 1442-1453, 1990

82. Reyes AD, Rubel EW, Spain WJ. *In vitro* analysis of optimal stimuli for phase-locking and time-delayed modulation of firing in avian nucleus laminaris neurons. J. Neurosci. 16: 993-1000, 1994

83. Rose JE, Gross NB, Geisler CD, Hind JE. Some neural mechanisms in inferior colliculus of cat which may be relevant to localization of a sound source. J Neurophysiol 29: 288-314, 1966

84. Rosowski JJ. Models of external- and middle-ear function. In: *Auditory Computation*, edited by H.L. Hawkins, T.A. McMullen, A.N. Popper, and R.R. Fay. New York: Springer-Verlag, 1996

85. Rothman JS, Young ED, Manis PB. Convergence of auditory nerve fibers onto bushy cells in the ventral cochlear nucleus: implications of a computational model. J. Neurophysiol. 70: 2562-2583, 1993

86. Ruggero MA and Semple MN, "Acoustics, Physiological", Encyclopedia of Applied Physics, vol. 1, 1991

87. Sanes DH, Malone BJ, Semple MN. Role of synaptic inhibition in processing of dynamic binaural level stimuli. J. Neurosci. 18: 794-803, 1998

88. Shackleton TM, McAlpine D, Palmer AR. Modelling convergent input onto interaural-delay-sensitive inferior colliculus neurones. Hearing Res. 149: 199-215, 2000

89. Shaw EAG. Transformation of sound pressure level from the free field to the eardrum in the horizontal plane. J. Acoust. Soc. Am. 56: 1848-1861, 1974

90. Shepherd GM, editor. The Synaptic Organization of the Brain, 5th ed., Oxford University Press, 2004

91. Shneiderman A, Oliver DL. EM autoradiographic study of the projections from the dorsal nucleus of the lateral lemniscus: a possible source of inhibitory input to the inferior colliculus. J. Comp. Neurol. 286: 28-47, 1989

92. Siebert WM. Ranke revisited — a simple short-wave cochlear model, J. Acoust. Soc. Am. 56: 596-600, 1974

93. Sivaramakrishnan S, Oliver DL. Distinct K currents result in physiologically distinct cell types in the inferior colliculus of the rat. J. Neurosci. 21: 2861-2877, 2001

94. Smith GD, Cox CL, Sherman SM, Rinzel J. Spike-frequency adaptation in sinusoidally-driven thalamocortical relay neurons. Thalamus and related systems 1: 1-22, 2001

95. Spitzer MW and Semple MN. Interaural phase coding in auditory midbrain: influence of dynamic stimulus features. Science 254: 721-724, 1991

96. Spitzer MW and Semple MN. Responses of inferior colliculus neurons to time-varying interaural phase disparity: effects of shifting the locus of virtual motion. J. Neurophysiol. 69: 1245-1263, 1993

97. Spitzer MW and Semple MN. Neurons sensitive to interaural phase disparity in gerbil superior olive: diverse monaural and temporal response properties. J. Neurophysiol. 73: 1668-1690, 1995

98. Spitzer MW and Semple MN. Transformation of binaural response properties in the ascending pathway: influence of time-varying interaural phase disparity. J. Neurophysiol. 80: 3062-3076, 1998

99. Steele CR. Behaviour of the basilar membrane with pure-tone excitation, J. Acoust. Soc. Am. 55, 148-162, 1974

100. Steele CR, Taber LA. Comparison of wkb calculations and experimental results for three-dimensional cochlear models, J. Acoust. Soc. Am. 65, 1007-1018, 1979

101. Steele CR, Zais JG. Effect of coiling in a cochlear model, J. Acoust. Soc. Am. 77: 1849-1852, 1985

102. Stevens SS and Newman EB. The localization of actual sources of sound. Am. J. Psychol. 48: 297-306, 1936

103. Sun Q, Gan RZ, Chang KH, Dormer KJ. Computer-integrated finite element modeling of human middle ear. Biomech. Model Mechanobiol. 1: 109-122, 2002

104. Tan Q, Carney LH. A phenomenological model for the responses of auditory-nerve fibers. II. Nonlinear tuning with a frequency glide. J. Acoust. Soc. Am. 114: 2007-2020, 2003

105. Viergever MA. Basilar membrane motion in a spiral shaped cochlea. J. Acoust. Soc. Am. 64: 1048-1053, 1978

106. Watkins AJ. The monaural perception of azimuth: a synthesis approach. In: Gatehouse RW (Ed.) Localization of sound: theory and applications. Amphora Press, 1982

107. Wernick JS, Starr A. Binaural interaction in superior olivary complex of cat – an analysis of field potentials evoked by binaural-beat stimuli. J. Neurophysiol. 31: 428, 1968

108. Wiener FM, Pfeiffer RR, Backus ASN. On the sound pressure transformation by the head and auditory meatus of the cat. Acta oto-laryngol (Stockh.) 61: 255-269, 1966

109. Wightman FL, and Kistler DJ. Sound localization. In *Human Psychophysics*, edited by W. A. Yost, A. N. Popper, and R. R. Fay. New York: Springer-Verlag, pp 155-192, 1993

110. Woodworth RS. Experimental psychology. New York: Holt, 1938

111. Xu Y, Collins LM. Predicting the threshold of single-pulse electrical stimuli using a stochastic auditory nerve model: the effects of noise. IEEE Trans Biomed Eng. 50: 825-835, 2003

112. Yin TCT and Chan JCK. Interaural time sensitivity in medial superior olive of cat. J. Neurophysiol. 64: 465-488, 1990

113. Yin TCT and Kuwada S. Binaural interaction in low frequency neurons in inferior colliculus of the cat. II. Effects of changing rate and direction of interaural phase. J. Neurophysiol. 50: 1000-1019, 1983a

114. Yin TCT and Kuwada S. Binaural interaction in low frequency neurons in inferior colliculus of the cat. III. Effects of changing frequency. J. Neurophysiol. 50: 1020-1042, 1983b

115. Yin TCT, Kuwada S, Sujaku Y. Interaural time sensitivity of high frequency neurons in the inferior colliculus. J. Acoust. Soc. Am. 76: 1401-1410, 1984

116. Yost WA. Fundamentals of hearing: an introduction, 4th ed. Academic Press, 2000

117. Young SR and Rubel EW. Frequency-specific projections of individual neurons in chick brainstem auditory nuclei. J. Neurosci. 3: 1373-1378, 1983

118. Yost WA and Hafter ER. Lateralization. In: Yost WA and Gourevitch G (Eds.) Directional Hearing, Springer-Verlag, 1987

119. Zhang X, Heinz MG, Bruce IC, Carney LH. A phenomenological model for the responses of auditory-nerve fibers: I. Nonlinear tuning with compression and suppression. J. Acoust. Soc. Am. 109: 648-670, 2001
120. Zweig G, Lipes R, Pierce JR. The cochlear compromise, J. Acoust. Soc. Am. 59: 975-982, 1976
121. Zwislocki JJ. Analysis of some auditory characteristics, in: R.R. Buck, R.D. Luce, E. Galanter (Eds.), Handbook of Mathematical Psychology, Wiley, New York, 1965.
122. http://hwr.nici.kun.nl/~miami/taxonomy/node20.html

Index

action potential, 7
auditory nerve, 124
auditory system, 108
axon, 6

basal ganglia, 61
basilar membrane, 115
bifurcation diagram, 25
binaural beats, 156
bistability, 26, 97
bursting oscillations, 43
bushy cells, 135

cable equation, 15
cell body, 6
cochlea, 115
cochlear afferent neurons, 116
cochlear nuclei, 130, 141
coupled cells, 50
cytoplasm, 2
cytoskeleton, 3

decibels, 112
dendrites, 6
depolarization, 7
depolarizing, 41

electrical circuits, 9
excitability, 41
excitatory, 96
excitatory synapses, 53

FitzHugh-Nagumo equations, 47
frequency tuning, 125

global bifurcations, 33

hair cells, 115
Hodgkin-Huxley equations, 15, 38
Hopf bifurcation, 24, 32
hyperpolarization, 7
hyperpolarizing, 41

inferior colliculus, 150
inhibitory, 96
inhibitory synapses, 57
inner ear, 115
innervation, 132
integrate-and-fire model, 72
intensity tuning, 125
interaural phase difference, 145
ion channels, 4

linear arrays, 86
local bifurcations, 31

midbrain, 150
middle ear, 114
Morris-Lecar, 147
Morris-Lecar equations, 39
multipolar cells, 136

Nernst equation, 5
nerve cells, 6
nerve terminals, 6
neural networks, 91
neuron, 37

octopus cells, 138

oscillatory networks, 100
outer ear, 113

Parkinson's disease, 62
peripheral auditory system, 117
perturbation theory, 34
phase locking, 126
phase models, 80
phase plane, 28
phase-resetting curve, 74
phaselocked state, 84
pitch chroma, 112
pitch height, 112
pitchfork bifurcation, 26
post-inhibitory rebound, 41
presynaptic terminals, 6
pyramidal cells, 139

Rall Theory, 18

refractory period, 8
rotation number, 76

saddle-node bifurcation, 25
singular perturbation, 35
slow synapses, 91
sodium-potassium pump, 4
soma, 6
sound localization, 144
spatial Hodgkin-Huxley system, 18
spiral waves, 90
superior olive, 142
synaptic coupling, 50
synaptic inputs, 154

thalamus, 161
transcritical bifurcation, 25
Turing-type instabilities, 101

4. For evaluation purposes, manuscripts may be submitted in print or electronic form (print form is still preferred by most referees), in the latter case preferably as pdf- or zipped ps-files. Lecture Notes volumes are, as a rule, printed digitally from the authors' files. To ensure best results, authors are asked to use the LaTeX2e style files available from Springer's web-pages at:

   ftp://ftp.springer.de/pub/tex/latex/mathegl/mono/ (for monographs) and
   ftp://ftp.springer.de/pub/tex/latex/mathegl/mult/ (for summer schools/tutorials).

   Style files for other TeX-versions, and additional technical instructions, if necessary, are available on request from lnm@springer-sbm.com.

   Careful preparation of the manuscripts will help keep production time short besides ensuring satisfactory appearance of the finished book in print and online. After acceptance of the manuscript authors will be asked to prepare the final LaTeX source files (and also the corresponding dvi-, pdf- or zipped ps-files) together with the final printout made from these files. The LaTeX source files are essential for producing the full-text online version of the book

   (http://www.springerlink.com/openurl.asp?genre=journal&issn=0075-8434).

   The actual production of a Lecture Notes volume takes approximately 8 weeks.

5. Authors receive a total of 50 free copies of their volume, but no royalties. They are entitled to a discount of 33.3 % on the price of Springer books purchased for their personal use, if ordering directly from Springer.

6. Commitment to publish is made by letter of intent rather than by signing a formal contract. Springer-Verlag secures the copyright for each volume. Authors are free to reuse material contained in their LNM volumes in later publications: A brief written (or e-mail) request for formal permission is sufficient.

Addresses:

Professor J.-M. Morel, CMLA,
École Normale Supérieure de Cachan,
61 Avenue du Président Wilson, 94235 Cachan Cedex, France
E-mail: Jean-Michel.Morel@cmla.ens-cachan.fr

Professor F. Takens, Mathematisch Instituut,
Rijksuniversiteit Groningen, Postbus 800,
9700 AV Groningen, The Netherlands
E-mail: F.Takens@math.rug.nl

Professor B. Teissier, Institut Mathématique de Jussieu,
UMR 7586 du CNRS, Équipe "Géométrie et Dynamique",
175 rue du Chevaleret
75013 Paris, France
E-mail: teissier@math.jussieu.fr

Springer-Verlag, Mathematics Editorial I, Tiergartenstr. 17,
69121 Heidelberg, Germany,
Tel.: +49 (6221) 487-8410
Fax: +49 (6221) 487-8355
E-mail: lnm@springer-sbm.com

GPSR Compliance

The European Union's (EU) General Product Safety Regulation (GPSR) is a set of rules that requires consumer products to be safe and our obligations to ensure this.

If you have any concerns about our products, you can contact us on ProductSafety@springernature.com

In case Publisher is established outside the EU, the EU authorized representative is:

Springer Nature Customer Service Center GmbH
Europaplatz 3
69115 Heidelberg, Germany

Batch number: 09490862

Printed by Printforce, the Netherlands